# Ingenious Inventions

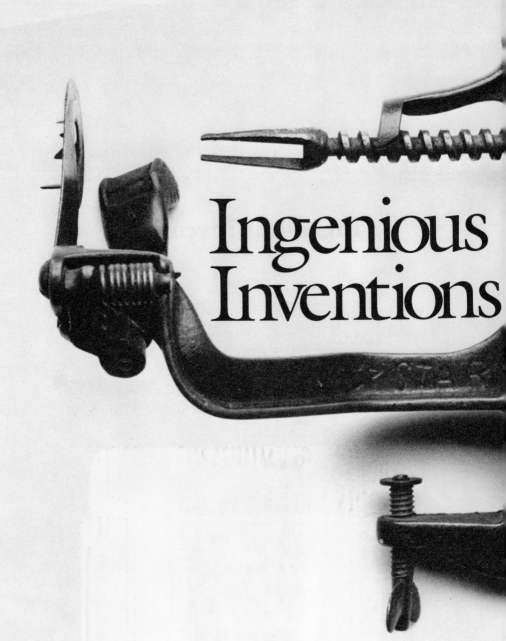

# Ingenious Inventions

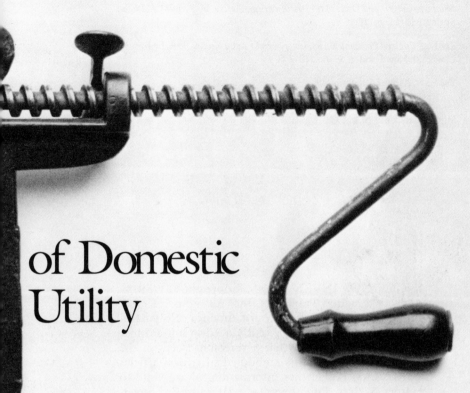

# of Domestic
# Utility

by Allen D. Bragdon

Text by Marcia J. Monbleau
Photographs by Ned Manter and Frank Foster

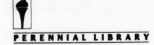

PERENNIAL LIBRARY

Harper & Row, Publishers, New York
Grand Rapids, Philadelphia, St. Louis, San Francisco
London, Singapore, Sydney, Tokyo, Toronto

The hints were selected from the following sources:

*Dr. Chase's Recipes; or, Information for Everybody*, Ann Arbor, Mich., 1866

*Manners, Culture and Dress of the Best American Society*, by Richard A. Wells, Springfield, Mass., 1890

*The New England Economical Housekeeper and Family Receipt Book*, by Mrs. E. A. Howland, New London, Conn., 1847

*The Sears, Roebuck Catalogue*, 1902

*Household Discoveries, An Encyclopaedia of Practical Recipes and Processes*, by Sidney Morse, Petersbury, N.Y., 1908

FIRST EDITION

LIBRARY OF CONGRESS CATALOG CARD NUMBER 89-45087
ISBN 0-06-096386-7

89 90 91 92 93    10 9 8 7 6 5 4 3 2 1

# Acknowledgments

We are grateful to those who made available the ingenious inventions in their collections and gave us permission to photograph them:

The Schoolhouse Museum and Swift-Daley House of the Eastham Historical Society, Eastham, MA; Sheila and Butch Frazier, Liberty Lore Antiques, Santuit, MA; Gloria Swanson of Tom Cardaroplis Antiques, 608 Route 6A, Dennis, MA; The Centerville Historical Society Museum, Cape Cod, MA; B. Kelsey Atkins, West Yarmouth, MA; The Whaling Museum, New Bedford, MA; Sharon Riedell; Rudolf Backlund; Marcy and Michael Pumphret; Marion Marsh; Nancy and Carl Clapp; Marcia J. Monbleau; Allen D. Bragdon.

Text researched and written by Marcia J. Monbleau. Page design and mechanicals by Deborah Sarafin Davies. Cover by "For Art Sake" from a design by Carolyn Zellers. Photography by Ned Manter and Frank Foster. Text formatting and copyediting by Lisa Clark and Karen Ringnalda Altman.

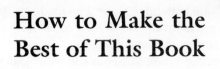

# How to Make the Best of This Book

**1. Test your ability to identify household utensils used in the "old" days before packaging design made it impossible to figure out what something was just by looking at it.**

To take that test, read this book, page by page, from front to back. The photo of each object appears first, its dimension shown in inches. You must turn the page to find its name and description. Occasionally that page may, instead, have a photo of something else used for the same purpose. In that case the name and description of both will appear on the next page.

**2. Catch glimpses of the daily lives of your great grandparents (or perhaps your own childhood if you already are a great grandparent).**

Americans have always been known for applying their inventive ingenuity to avoid repetitive hand labor or having to bend over. The tools and apparatus they devised in the past reveal that they spent a lot of time doing both. Even the labor-savers were hand-driven (or foot-driven, as in bootjacks or the remarkably simple utensil used to capture a hen free-ranging around the back yard when both of you knew it would soon be dinnertime).

**3. Pick up ideas that may be useful if the power and phones go off for a month or so.**

In those days people bought their necessities in bulk, converted them by hand, and they carried their lighting with them up the dark stairs to bed after dinner. Reprinted here are household hints, helps and cures related to each object described. These are reproduced from some old housekeeping books listed on the Acknowledgments page. It is interesting to note the extent of the skills and knowledge a woman had to master when the only household cleaning miracle was soap and there was no way to call the doctor.

**4. Sharpen your collector's eye. Many of these old curiosities are increasing in value faster than the national debt, even.**

Here are a few facts that could help you date some attic treasures of your own. Most utensils that have survived till today were made from the mid-1880s on. After the Civil War, America industrialized. Long-lasting utensils began to be manufactured from cast iron. Patent law encouraged manufacturers to stamp them with the patent date. Until then most of the population had brought their essential and valuable utensils with them from Europe. Or if they needed a new tool for a specialized task, they knocked one together as best they could from cheap and available materials—usually wood. Most were farmers, not toolmakers. Soon the contrivance broke or fell apart, and the discarded pieces rotted away or became part of something else. Since the country-of-origin law wasn't passed until 1896, a utensil that doesn't have a country's name on it was either handmade or probably dates from before then.

However you use this book, we all hope you have a good time with it.

Allen Bragdon
South Yarmouth, Mass.
August 1989

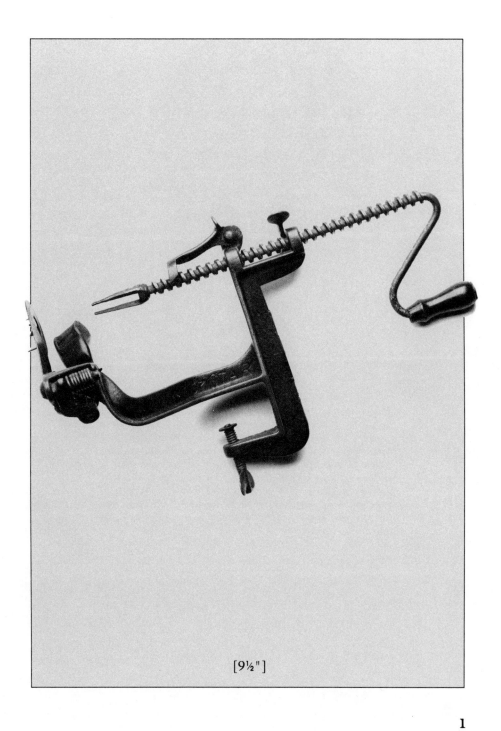

[9½"]

To dry or evaporate apples, peel and core them and cut in thin slices. Let the slices fall into cold water to prevent their rusting. Lay the slices on a large piece of cheese cloth and baste them to this by means of a darning needle and suitable cotton thread, taking a stitch through each slice, so that it will lie flat and keep in place. Suspend the cheese cloth out of doors by the four corners, high enough to be out of the reach of small animals. Spread another thickness of cheese cloth over the fruit and expose to direct sunlight. When sufficiently dry, store them in a dark place.

[ANSWER ON PAGE 4]

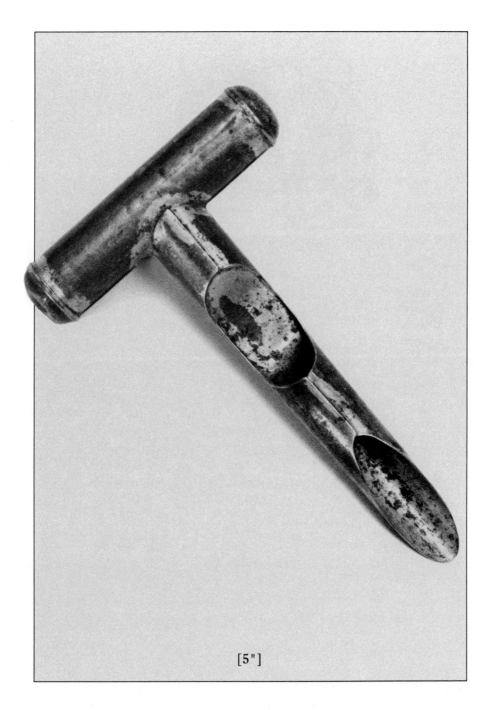

[5"]

## APPLE PEELER/CORER

While it might be a pleasant relaxation nowadays to sit on the back porch and peel apples for a pie or two, preparing a bountiful harvest of fruit for a winter's worth of eating used to be a monumental chore. When it was apple-picking time, they were bought by the bushel or barrel, cooked down for sauce, put up as jelly or dried, so they could be used months later. A barrel of apples meant a barrel of work.

Clamped to the kitchen table, this turn-of-the-century, turn-of-the-screw contraption performed three functions. The handle turns a corkscrewlike shaft with a three-pronged end to hold the apple in place. As the handle turns, it moves the apple toward three blades. One removes the peel, another slices the meat into a continuous spiral, and the third reams out the core. Before you could say "Johnny Appleseed," the apple was peeled, cored, and spiraled. Those spirals could then be hung from the ceiling until dried. You didn't need to go to all that effort if you wanted to make, say, baked apples. The little tin utensil just punched out the core.

[24" x 24"]

## BAYBERRY CRUSHER

In the earliest days of America, colonists discovered that the very choicest of candles could be made from the fruit of a low bush growing in sand dunes along the New England seashore. The small, grayish bayberry was picked, crushed in a device like the 1740 one shown here, and boiled. It had to be skimmed several times before the light, almost transparent green fat was sufficiently refined. Bayberry candles were highly prized, because so many berries were needed to make just one. Peter Kalm, a Swedish naturalist who visited America early in the 18th century, said that bayberries "grow abundantly and look as if flour had been strewed upon them. The wax is dearer than tallow but cheaper than beeswax. Such candles do not bend easily nor melt in summer. They burn better and without smoke, yielding an agreeable smell when extinguished."

To make bayberry soap, dissolve 3½ ounces of white potash in 1 pint of water, and add 1 pound of bayberry tallow. Boil slowly and stir until the mixture saponifies. Add 2 tablespoons of cold water containing a pinch of salt and boil 5 or 6 minutes longer. Remove from fire, and when it is cool, but before it sets, perfume by adding 5 or 6 drops of any essential oil, according to taste. This soap is valuable for all toilet purposes.

[12" sq]

7

## BOG SHOES

From about 1870 to 1920, these ingenious objects of wood, rope, and iron were used on workhorses when either land or weather conditions could not be permitted to interfere with the day's work. The pieces of wood were cut and given to a blacksmith, who turned the iron and attached it to the base, making shoes that prevented the horses from sinking into mud or muck. They were used, for example, when wood-cutting chores had to be done and the logging road was a mire, when a sugaring wagon had to be pulled along a muddy road made almost impassable by the spring thaw, and when men went into the lowland to cut and haul marsh grass. The ones shown here are 12 inches square.

*For horse ointment, 4 oz. resin, 4 oz. beeswax, 8 oz. lard and 2 oz. honey. Melt these slowly, gently bringing to a boil; and as it begins to boil, remove from fire and slowly add a little less than a pint of spirits of turpentine, stirring all the time. Cool. This is extraordinary for bruises, in flesh or hoof; broken knees, galled backs, bites, cracked heels. It is excellent to take fire out of burns or scalds in human flesh, also.*

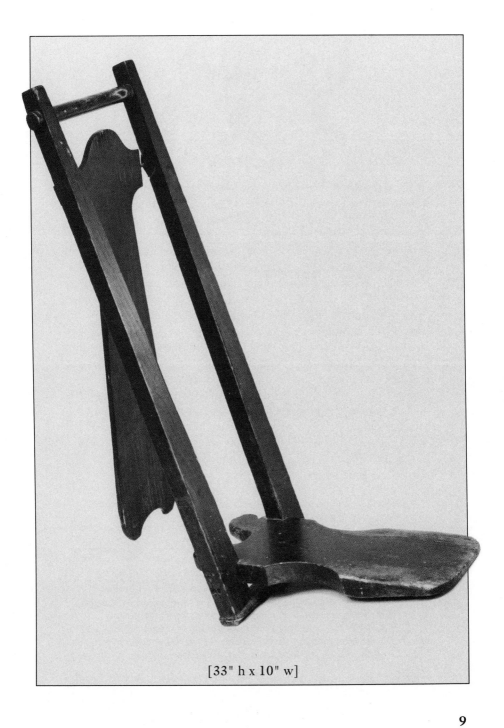

[33" h x 10" w]

## BOOTJACK

In the days when men wore high boots—either for military dress, riding, or work, pulling those boots on was one thing but getting them off, quite another. The jackboot worn by cavalry officers came up over the knee, so a man couldn't even bend his own leg enough to take off his boot. The fortunate ones had a manservant to help with the tugging. A farmer might corner a son or wife to help take off his work boots. Either way, it was a two-person job.

This 1740s bootjack, possibly of Shaker design, took the place of that second person. (Beginning in the late 18th century, the American communities of Shakers, a religious sect, made furniture and wooden utensils that have been greatly admired for their gracefulness, practicality, and simplicity of style.) The original bootjack was nothing more than a wedge of wood with a V-shaped notch on one side. Step on the near side of the wedge, hook the heel of your other boot in the notch, and pull, easing your foot out of the boot. This jack is more substantial. The swinging flap of wood secures the top of the boot as well, so it doesn't lift out of the notch, and the top rail gives the wearer something to hold onto during the whole process.

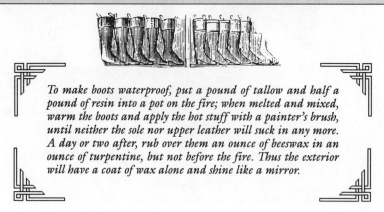

*To make boots waterproof, put a pound of tallow and half a pound of resin into a pot on the fire; when melted and mixed, warm the boots and apply the hot stuff with a painter's brush, until neither the sole nor upper leather will suck in any more. A day or two after, rub over them an ounce of beeswax in an ounce of turpentine, but not before the fire. Thus the exterior will have a coat of wax alone and shine like a mirror.*

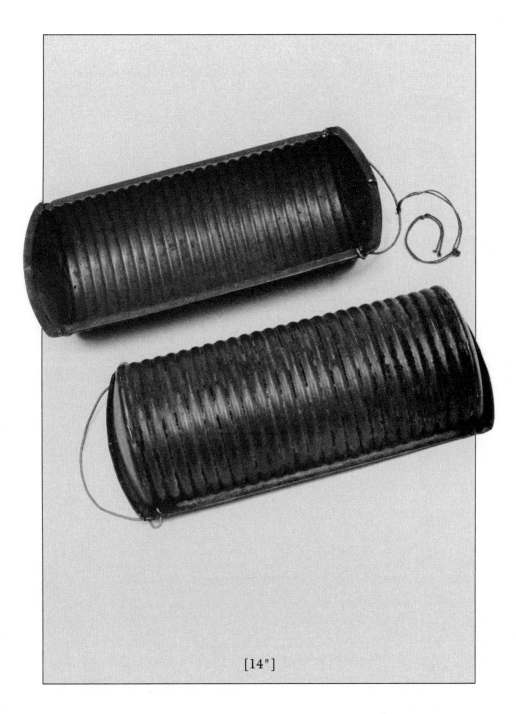

[14"]

11

## BREAD LOAF BAKER

This is a tin bread pan, circa 1900, with a seam on one side, a closure latch on the other. It guaranteed an evenly-cooked loaf, and the rippled edge provided that equal-sized slices could be cut.

In a day when precious little on the family table was store-bought, the woman of the house had to plan her week carefully to allow time for cooking chores—as well as the many others that were necessary to the proper running of a home. Monday was usually set aside as washday, and it was thought that the following day should be devoted to ironing, sewing, and mending. A third day was taken up with housecleaning. But one day a week was reserved for baking—cakes and pastries sometimes, bread always.

*The process of kneading has more to do with good bread than almost anything else. I have seen pieces of the same dough, raised in the same temperature, baked in the same oven, yield two entirely different qualities of bread. One loaf was molded by an energetic, strong-muscled girl whose kneading was so strenuous that all the life had been banged out of it. The other loaf was kneaded by a girl whose every movement was grace; she used her hands deftly, lightly and briskly. Her bread was as fine as bread could be made. It is not brute force that tells in kneading.*

[4", 5", 2¾" x 5"]

## BUTTER MOLDS

These small wooden molds were used throughout the 18th and 19th centuries, not out of necessity, but because of a desire to make more decorative—more attractive for special events—one of the most common accompaniments to any meal. They could be fashioned in many shapes, of every design and for particular holidays throughout the year. When good, rich cream was churned into butter, it was common to find a plain solid slab or hunk of it on the everyday table. But when the occasion demanded, new butter was pressed into these molds to create pretty, individual pats. The round ones here were handcarved in about 1800, the rectangular one in the 1880s.

*If buttermakers or dairymen will use only shallow pans for their milk—and the larger the surface, and the less the depth of the milk the better—then put into each pan, before straining, 1 quart of cold spring water to every 3 quarts of milk, they will find the cream will begin to rise immediately, and skim every 12 hours, the butter will be free from all strong taste arising from leaves or coarse pasturage.*

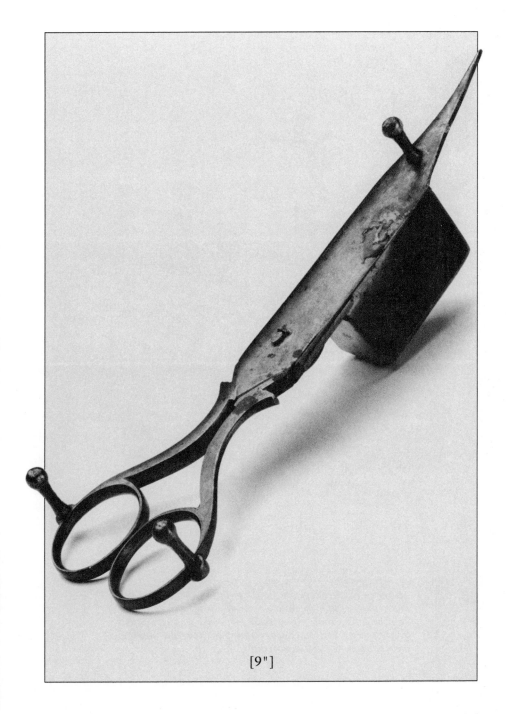

[9"]

## CANDLE SNUFFER

Before the advent of kerosene and gas lamps, before the bright miracle of electricity, light was carried from room to room at the end of a candle. Those candles—usually homemade—were tended carefully and not permitted to burn any longer than necessary. Beginning in about 1870, candle snuffers appeared here and there throughout the house. They operated on the same principle as a pair of scissors and served three purposes. First, they put out the flame. At the same time they both clipped or trimmed the wick and neatly caught that little bit of rubbish in the box-like receptacle. The burned or charred part of the wick is called the "snuff." When the snuff was trimmed off, the candle would burn without smoking.

*To light a candle, apply the match to the side of the wick and not to the top. To blow out a candle, if a snuffer is not available, hold the candle higher than the mouth and blow it out by an upward instead of a downward air current. This will prevent the wick from smoldering.*

[3½"]

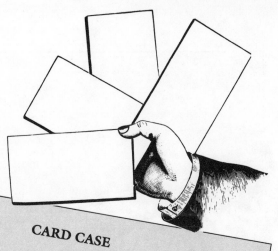

## CARD CASE

In the late 19th and early 20th centuries—before homes were connected by telephone—people could get in touch with each other only by writing or visiting, and there were certain rules of etiquette attached to the latter. Visits usually were planned in advance, mid-afternoon was the appropriate time of day (never in the morning when the lady of the house was occupied with domestic duties), and, unless otherwise specified, 15 minutes was thought to be the proper length of a call. Such courteous behavior meant that guests did not barge in at inopportune moments, nor did they dawdle, waiting to be invited to dinner.

Good manners demanded that a card be presented, to either arrange a call or announce the visitor's arrival. Both ladies and gentlemen had cards imprinted with their names. This 1860s sterling silver calling card case, carried in a vest pocket, is slightly curved to fit a gentleman's form.

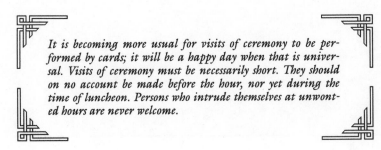

*It is becoming more usual for visits of ceremony to be performed by cards; it will be a happy day when that is universal. Visits of ceremony must be necessarily short. They should on no account be made before the hour, nor yet during the time of luncheon. Persons who intrude themselves at unwonted hours are never welcome.*

18

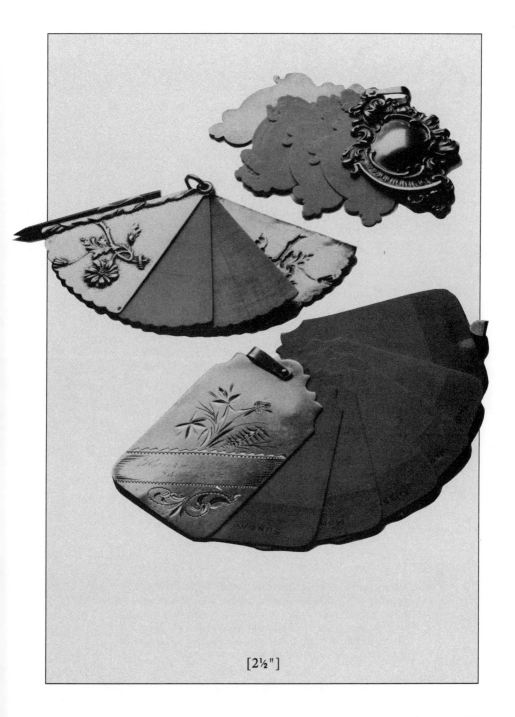

[2½"]

## CARTE DE BAL

Each of these elegant objects is about 2½ inches high, and consists of a silver outer leaf that opens like a fan to reveal several thin pages or leaves of ivory. One of them is engraved with a man's name, dated June 1895, and contains seven leaves, each of which is stamped with a day of the week. The clip at the top indicates that the case was attached to a watch chain and carried in the vest pocket. These may have been used in one of two ways. Certainly the one with the days noted on it served as a memo pad to keep track of business and social engagements. But each is also a "carte de bal," or dance card. It was proper for a gentleman to ask a lady for a particular dance during a social gathering. Women and men alike carried dance cards. They wrote down the name of their partner for each dance to avoid committing a grave social error. At the end of a dance, pencil marks on the ivory pages were easily wiped off.

*The master of the house should see that all the ladies dance; he should take notice, particularly, of those who seem to serve as drapery to the walls of the ball-room (or wall-flowers, as the familiar expression is) and should see that they are invited to dance. But he must do this wholly unperceived, in order not to wound the self-esteem of the unfortunate ladies.*

[14"]

*For cherry jam, weigh the fruit before stoning and allow ½ pound of sugar for each pound. For every six pounds of cherries, allow 1 pint of red currant juice, and for every pint of juice, allow 1 pound of sugar. Stone the cherries and boil them until almost all juice is gone. Add the sugar and currant juice. Boil until it jells, skim, stirring well.*

[ANSWER ON PAGE 24]

[10¼" l x 6⅝" w]

## CHERRY STONERS

These two cast-iron utensils are different versions of the same thing—and much like the one pictured in Sears, Roebuck's 1902 catalogue that was priced at 63 cents. Feed the cherries into the machine. Turn the handle, forcing the fruit through a narrow channel and squeezing out the stone. They were made to do "with rapidity" what otherwise would have been an everlasting job for a housewife or family cook setting out to make jam, brandy, sauce, or pies. Kitchen aids like this saved time and effort when the produce of summer had to be preserved for winter use. Fresh fruits and vegetables were not shipped from place to place as they are today. They were available in any given part of the country only when in season there. A family could not enjoy fruit in the spring if it hadn't been put up the previous summer.

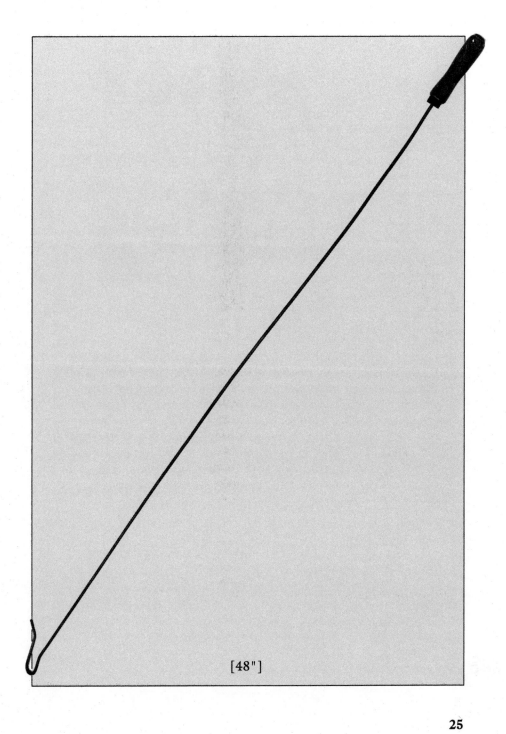

[48"]

## CHICKEN CATCHER

Here is further proof that necessity is the mother of invention. Perfectly simple, with a handle at one end and small hook-turn in the metal at the other, this tool was created, no doubt, so a farmer wouldn't have to run around like a chicken with its head cut off. When someone went to the coop or barnyard to get a hen for dinner—or several for market—it was no easy chore. Perhaps the bird had an inkling of what was to happen, but she didn't stand around to find out. It was hard to grab a chicken. With this device, you could stand a fair distance away and snag supper by the legs.

[10" h x 5" dia.]

## CHICKEN WATERER

This oddly shaped, highly glazed farmyard utensil from the early 1900s is pointed, and there's a definite point to that point. The base is open, and there is one more very small hole just above it. The container was turned upside down and filled with water. A shallow pan was then clapped firmly over the opening and the whole shebang was flipped over quickly and placed on the ground. Water flowed out of the hole to fill the pan, and the chickens could quench their thirst. Unlike other watering devices, this one had a practical advantage because of its shape. The point and smooth surface prevented the birds from perching on top, from which location they would otherwise both decorate the container and foul the water.

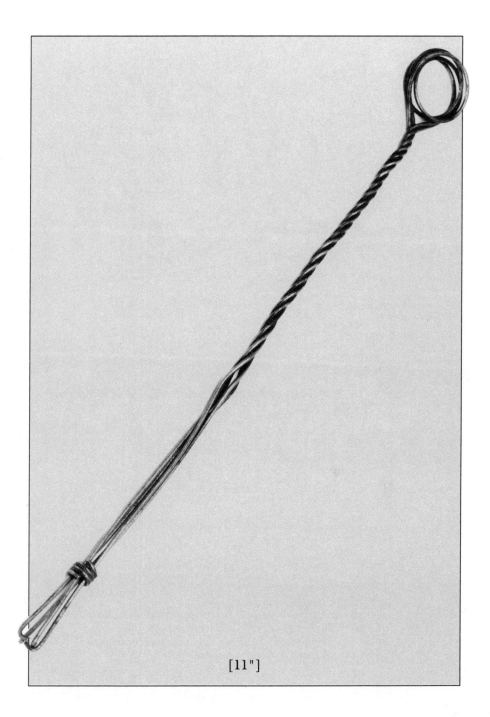

[11"]

## CHIMNEY CLEANER

When kerosene or oil lamps provided all lighting in the home, the careful housekeeper set aside time each day to replace the fuel, trim the wicks, and wash or wipe out the narrow glass chimneys, because dirty lamps gave off poor light. (This chore often was assigned to an older child in the family.)

This little wire tool, which sold for a nickel in 1900, was made to clean chimneys, most of which were too small to reach into. Open the end by sliding back the small ring, insert a rag, slide back the ring—making fast the rag—then feed the tool into the sooty chimney. The one pictured here is from the 1870s.

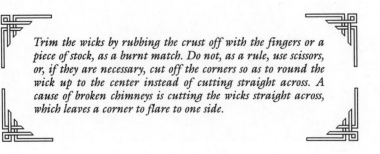

*Trim the wicks by rubbing the crust off with the fingers or a piece of stock, as a burnt match. Do not, as a rule, use scissors, or, if they are necessary, cut off the corners so as to round the wick up to the center instead of cutting straight across. A cause of broken chimneys is cutting the wicks straight across, which leaves a corner to flare to one side.*

[48" x 60"]

## CIDER PRESS

"Cider is to be found in every house," said an 18th-century New York farmer. "Our cider affords the degree of exhilaration with which we are satisfied."

Two hundred years ago, apple cider was called the "vin du pays" of America. The ingredients for this "wine of the country" were easy to come by, and the cider-making process was a simple one. This was a good homemade drink, and some thought it even better after it had fermented and become mildly alcoholic or "hard."

In the 1840 presidential campaign, William Henry Harrison was made light of as a simple frontiersman. But the incumbent president, Martin Van Buren, preferred champagne, which was eventually used as a weapon by the opposition in the following verse:

> "Let Van from his coolers of silver drink wine,
> And lounge on his cushioned settee.
> Our man on his buckeye bench can recline,
> Content with hard cider is he."

Cider-making began at fall harvest time. The fruit was ripe, according to a Shaker apple-grower, "when at 30 feet distance you catch the fragrant apple aroma." Apples were picked, then allowed to rest a few days, in order to gather juice and flavor. Next, they were crushed to a pulp. Finally, the juice was "expressed" from the fruit by forcing it through a press like the 1890 model pictured here. The handle on the long corkscrew rod was turned, and the liquid literally squeezed or pressed out of the pulp into a bucket below. This press measures four feet by five feet.

*"Cider and doughnuts make old people's tales and old jokes sound fresh...and enchanting, and juggle an evening away before you know what went with the time."*

*Mark Twain*

*To prepare cider for medicine—to each barrel of cider just pressed from ripe, sour apples, not watered, take mustard seed, unground, 1 lb; isinglass 1 oz; alum pulverized 1 oz; put all into the barrel, leave the bung out and shake or stir once a day for four days, then take new milk one qt, and half a dozen eggs, beat well together, and put them into the cider and stir again for 2 days; then let it settle until clear.*

[12"]

## CORSET BUSK

The world of ladies' fashions didn't begin to, shall we say, loosen up until early in the 20th century. Before then, the fashionable hourglass figure was shaped by a system of wires, bones, and laces so constricting that taking a deep breath must have been nearly impossible. A lady getting dressed in her finest could not do so without help. Fabric corsets, reinforced with bone stays, had to be laced up in the front or the back—either way requiring an extra pair of hands. The "main stay," or "busk," was larger than the others, and this was inserted vertically in the front of the corset. It was flat, about an inch-and-a-half wide and from 8 to 12 inches long depending on the height of the woman.

This corset stay, or busk, was made in the 1820s and was, more than likely, a sailor's gift to his wife or betrothed. It is a piece of baleen, decorated with scrimshaw—the art of etching on bone, shells, or ivory. Sailors on long whaling voyages drew pictures reflecting either memories of home or adventures at sea. Scrimshaw pieces rarely contained any words, because most sailors couldn't read or write. As early as the 16th century, a corset busk was considered a very intimate gift.

*"Accept, Dear Girl, this busk from me;*
*Carved by my humble hand.*
*I took it from a Sparm Whale's jaw,*
*One thousand miles from land!*
*In many a gale, has been the whale,*
*In which this bone did rest.*
*His time is past, his bone at last*
*Must now support thy brest."*

Clifford Ashley
New Bedford whaleman, artist, and author
1881–1947

34

[18" x 24"]

## CRANBERRY SCOOPS

These wooden scoops were used into the mid-20th century, when machines finally were developed to mechanize the harvesting of cranberries grown in Massachusetts, New Jersey, Wisconsin, Washington, and Oregon. Until that modernization, workers bent over, or knelt on the ground, and literally scooped the fruit from the low-growing vines, filling the container and then emptying it into a wheelbarrow or barrel. It was a long and back-breaking process. These cranberry scoops were hand-crafted—most of them custom-made for the individual picker—and could be as small as 18 teeth or as large as 36 teeth wide. On Cape Cod in Massachusetts, they were often made by fishermen during the winter months when it was too cold to be out on the water.

The larger of the two, a "rocker scoop," was used in just the manner its name implies—rocked back and forth along the cranberry bog, lifting the berry and breaking it off from the vine. The 12-inch "snapper scoop," with a spring closure that trapped the fruit, was used on new, tender vines, and, because of its size, was also used by children who worked alongside the adults during the fall harvest. The one here was made in 1920.

Most cranberry bogs are 10 to 12 acres, and the largest one in the world—once covering more than 300 acres—is on Nantucket Island.

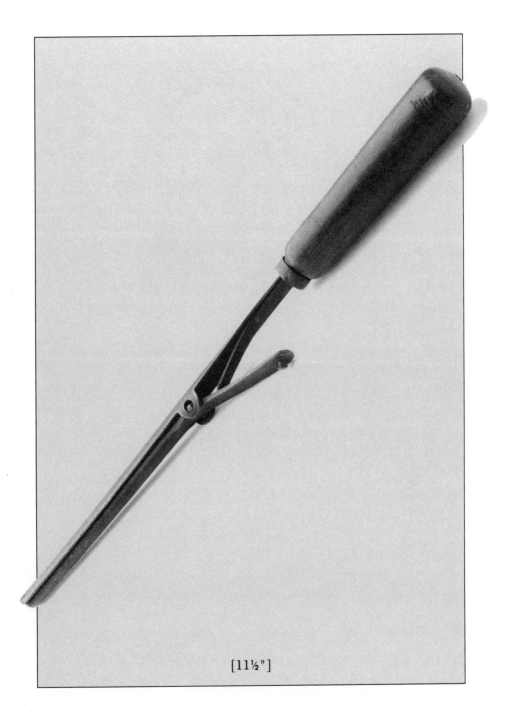

[11½"]

Perhaps the best of all shampoos is the yolk of an egg beaten up with a pint of soft warm water. Apply at once and rinse off with castile or other hard white soap.

To keep hair in curl, take a few quince seeds, boil them in water and add perfuming. Wet the hair with this, and it will hold a curl. It also is good to keep the hair in place when going out in the wind.

[ANSWER ON PAGE 40]

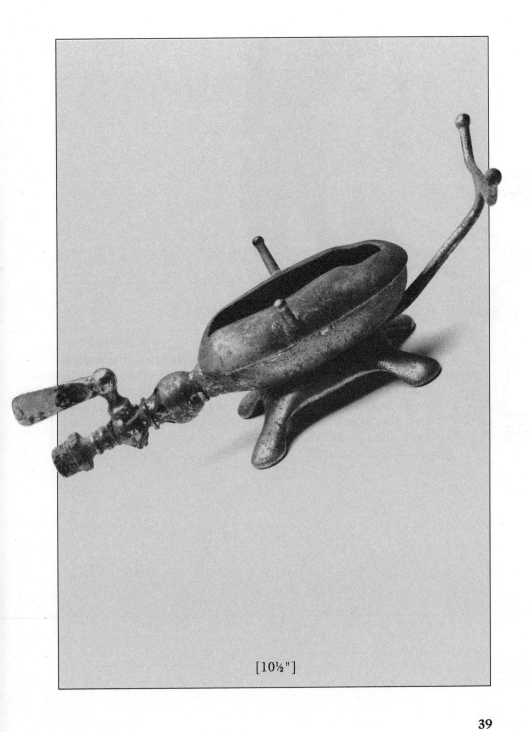

[10½"]

## CURLING IRON

A lady wishing to be in fashion with a curl or pinched-in wave repaired to her curling iron. It might be called the first cordless hair curler, because although modern versions exist, this early (circa 1830) model used the kitchen wood stove, alcohol heater, or gas jet as its source of heat. It is among the dozens of household items that were devised and in everyday use long before anyone ever heard the word "electricity."

[11"]

## DROP SPINDLE

When a spinning wheel was not available, raw wool was twisted into weavable yarn with a drop spindle. Before the yarn could reach the spindle stage, however, the raw, dirty fleece had to be washed in cool water (hot water would remove the lanolin from the fibers and make them less waterproof). Then the wool had to be fluffed and cleaned by using "cards"—wooden, paddlelike frames containing rows of metal teeth. You hold a card in each hand and gently comb the wool from one to the other, repeating the process several times until the wool is airy and all short bits have been removed. After carding, the wool is ready to spin. To use the drop spindle, twist a few bits of the fluffy "rolag" into a short length of yarn and tie it to the top pointed end of the spindle. Rest the rolag on the back of your left hand. As you keep the spindle twisting in the air ("spin"-dle, get it?) you keep adding bits of wool from the rolag and twisting it into yarn. When the yarn is so long the spindle touches the ground, wind up the yarn around the spindle to store it and play it again, Sam, from the top.

*Woolen goods must not be soaked, boiled, scalded or wrung out by twisting. They must not be dried near a hot fire. The fibers of wool are hooked and curled, and when they are crushed together by rubbing they form knots, which thicken the fiber and shrink it in both dimensions. This is one of the principal causes of the shrinking that is so feared. Or the expansion and contraction caused by alternate heat and cold may cause the fibers to interlace.*

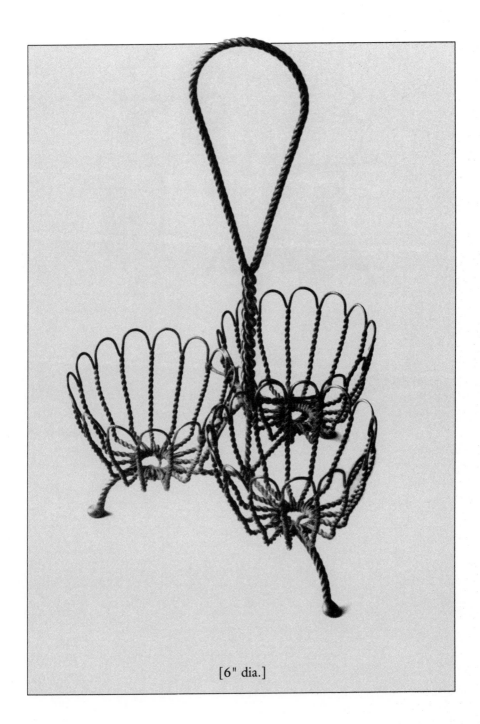

[6" dia.]

## EGG BOILER

Eggs that had been kept in some sort of refrigeration were placed in the basket to be set in warm water, so the change in temperature would not crack the shells. The basket of eggs was then put into boiling water for the desired amount of time.

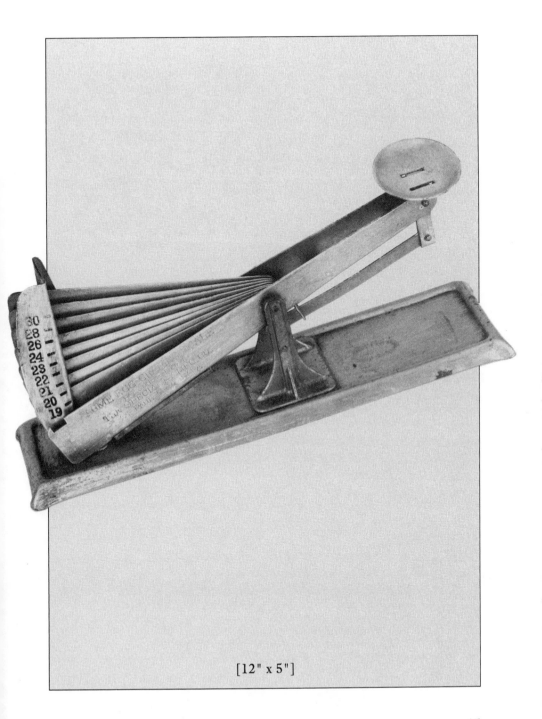

[12" x 5"]

## EGG SCALE

This 12-inch tin scale, circa 1924, was found on farms where there was a businesslike hen house—not one for just family use. Eggs were sold by the dozen or half-dozen, obviously, but also by size, and that size—or grade—was best determined by weight. Although farmers usually took most of their milk and eggs to markets and dairies, it wasn't uncommon for them to stop along the way and make deliveries to friends and neighbors who were partial to fresh milk and eggs with perhaps a feather or two stuck to the shell.

*To preserve eggs, take a keg or pail, cover the bottom with a half-inch of salt and set your eggs close together, on the small end; sprinkle them with salt so as to cover them entirely, and then put down another layer and cover with salt, till your keg is full; cover it tight, and put it where they will not freeze, and they will keep fresh a good year or longer.*

[can 3" dia.]    [rod 12" long]

## FIRE LIGHTER

In the days before central heating, when homes were warmed by wood stoves and fireplaces, fire lighters like this brass one copied from an 1870s original could be found on the hearth. The lighter itself was a piece of stone on a thin metal rod. A member of the household filled the container with kerosene and left the stone to soak in it. After laying the fire with bits of kindling or wood shavings, he or she thrust the fire lighter in among them and put a match to it. Because the stone was saturated with kerosene, it stayed ablaze long enough for the wood to ignite. After the fire had got going, the starter was retrieved, left on the hearth to cool, then placed back in the container where it would be nicely soaked in time for the next fire.

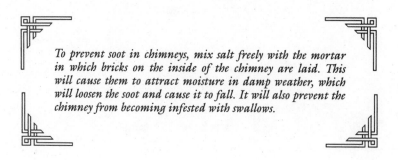

*To prevent soot in chimneys, mix salt freely with the mortar in which bricks on the inside of the chimney are laid. This will cause them to attract moisture in damp weather, which will loosen the soot and cause it to fall. It will also prevent the chimney from becoming infested with swallows.*

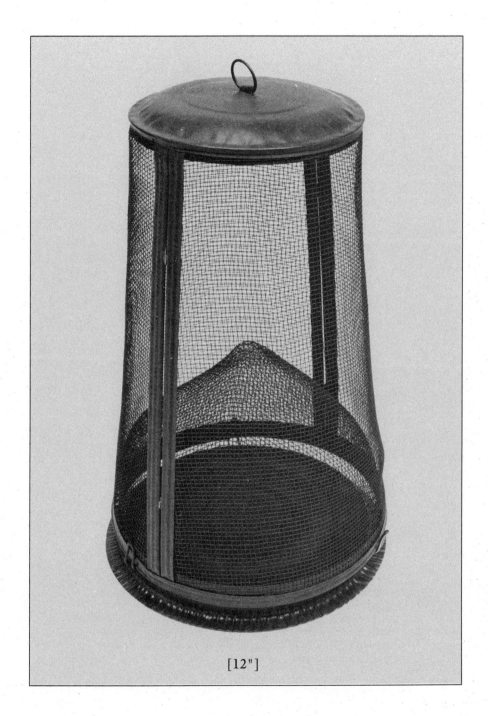

[12"]

## FLY CATCHER

Although not a household necessity, this deceptively simple wire object from the 1860s certainly made life less annoying wherever it was put to use. Of a fine mesh, it has a hook on top, a removable bottom, and a narrow opening all around the base. Put a small dish of honey or sugar inside and hang it up in the house or barn. Flies go in through the bottom opening in search of something sweet, and because they must fly upward to take off, they never figure out how to escape.

*Flies are said to abhor sweet clover. Place in bags made of mosquito netting and hang about the room. Or sprinkle oil of sassafras or oil of laurel. The latter has been used by the butchers of Geneva from time immemorial. Or use oil of lavender or lavender buds. Or boil onion in a quantity of water and wash picture frames, moldings and woodwork with the concoction. To prevent flies from settling on windows, wash them in water containing kerosene and wipe with a rag moistened with kerosene.*

[7" dia.]

## FOOD MILL

This old-time implement is still being manufactured, in spite of today's electric food processors. When it came time for the harvest to be put up for winter consumption, the simple metal device mashed everything from carrots to squash. It strained food for children and invalids and was used to prepare jellies and preserves. When making applesauce, just quarter the apples and cook them down. When you put the cooked fruit in this food mill and turn the paddle, the peel, seeds, and stems all are trapped by the strainer, and only the finest sauce goes through to the bowl below.

In a 1937 advertisement, the manufacturer said that the mill would juice a bushel of tomatoes in 15 minutes and convert a bushel of apples into sauce in a half-hour.

*To keep apples for winter use, put them in casks or bins, in layers well covered with dry sand, each layer being covered. This preserves them from air, from moisture and from frost; it prevents their perishing by their own perspiration, their moisture being absorbed by the sand; at the same time it preserves the flavor of the apples and prevents their wilting. Any kind of sand will answer, but it must be perfectly dry.*

[11½" w, 8" d, 6" h]

## FOOT WARMER

In the days when heat was produced by fireplaces, stoves, and hard work rather than by the twist of a thermostat, it was the subject of considerable attention, both in and out of the home. The horse and buggy was not noted for its heating system, nor were most places of public gathering, so the family often carried its own supply of warmth when traveling or in church.

This early tin foot warmer, circa 1780, filled with hot coals, was carried from here to there, and it helped an individual ward off a chill during a short ride or a long sermon.

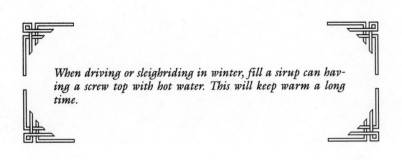

*When driving or sleighriding in winter, fill a sirup can having a screw top with hot water. This will keep warm a long time.*

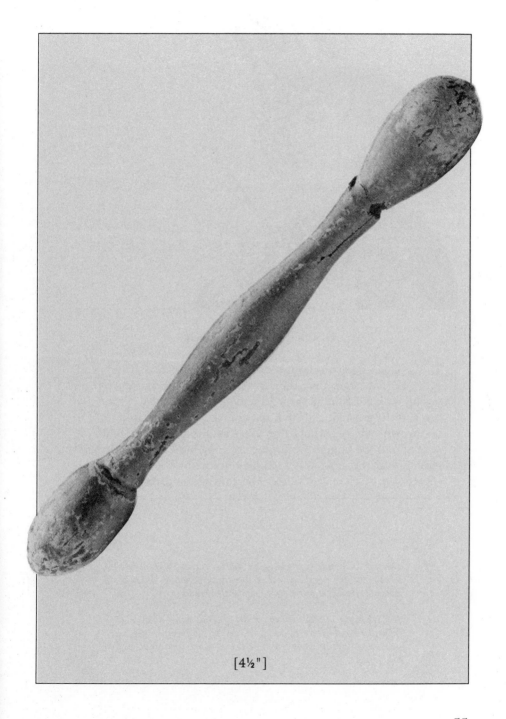

[4½"]

## GLOVE DARNING EGG

This wooden object—about six inches long, rounded at each end, with one end larger than the other—was found in every woman's sewing basket. Ladies never left the house without their gloves and hat, and the fact that a glove had a hole in it was no reason for it to be thrown out. A darning egg like this one from the 1880s was pushed into the finger (the larger end being used for a gentleman's glove), and the hole was mended—sometimes almost invisibly (depending on the talent of the woman with the darning needle).

*Instead of mending gloves with silk to match them as is the usual custom, try good cotton thread the color of the gloves. You will find it not so noticeable as silk thread*

*Jewelry is out of place in any of the errands which take a lady from her home in the morning. Lisle thread gloves in summer and cloth ones in winter will be found more serviceable than kid ones.*

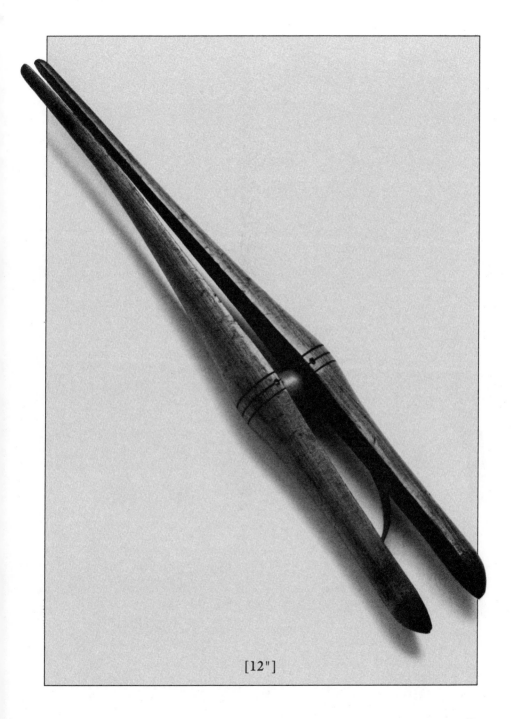

[12"]

# GLOVE DRYERS/STRETCHERS

Until very recently gloves were a proper part of any lady's attire, whether leaving home to pay a call, shop, go to church, or attend any social gathering. Such incessant use meant there was a constant need to wash them. Most often, a woman would clean her fine kid gloves by putting them on and washing them (as if washing her hands) with either soap and water or spirits of turpentine. After doing so—and without wringing them—she would use a wooden glove-stretcher like this one from the 1880s. Insert the smaller end into a glove finger and, by squeezing the grip, expand the stretcher to ease and reshape the leather. Then hang the gloves in the sun to dry and get rid of the turpentine smell.

Kid gloves could also be turned inside out, brushed with cold cream or any melted mixture of oil, lard, or other unguent, turned right side out again and worn to bed overnight to help soften skin or cure chapped hands.

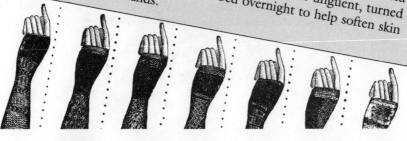

To wash kid gloves, draw the gloves on the hands, dip a cloth in skim milk and wipe them on a cloth as if on a towel. Or draw a glove on one hand and, with the other, dip a piece of flannel in new milk, rub on castile soap or any good hard white soap and rub the soiled glove lightly.

For an ointment, melt together 3 drams of gum camphor, 3 drams of beeswax, 2 ounces of olive oil. Apply at night and wear chamois-skin gloves to bed.

[4" dia.]

## HAND GRENADE

This blue glass bottle is beautiful enough to be placed on a what-not shelf or sunny windowsill, but it is far from a frivolous decoration. When home heat was provided by a wood fire in both living and bedrooms, fire-fighting equipment of any sophistication—horse-drawn water trucks—was limited to urban areas, and country firefighters were, for the most part, good neighbors with wooden buckets. In either case, responding to an emergency took time—often too much time.

This is an 1870s "Harden Hand Grenade," and it sat on the mantel. If a fire got out of control, the grenade—filled with a special chemical liquid—was thrown into the fireplace. The glass smashed, and the flames were doused. Use of these was stopped, however, when it was discovered that fumes from the liquid were harmful to people.

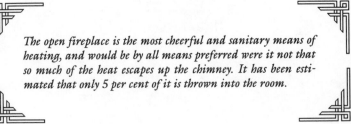

*The open fireplace is the most cheerful and sanitary means of heating, and would be by all means preferred were it not that so much of the heat escapes up the chimney. It has been estimated that only 5 per cent of it is thrown into the room.*

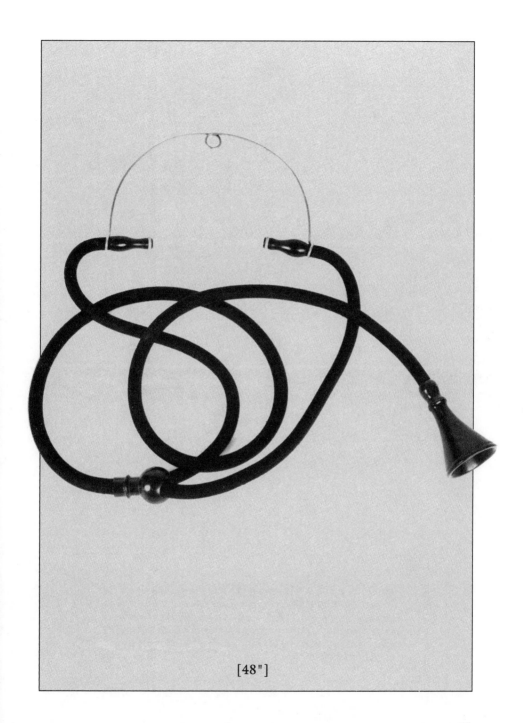

[48"]

## HEARING AID

While this might look very much like a section of garden hose, it was not intended to carry water. Rather, it was designed in 1910 to provide a measure of relief for the hard of hearing at a time when such simple devices were state-of-the-art in the medical world. It works on a very basic principle: If you speak into a piece of pipe, the sound coming out the other end will be amplified. Carrying around a length of pipe is highly impractical, but the same theory applies to these four feet of rubber tubing. The hard-of-hearing person could carry the device with him, hold one end to his ear, and point the other toward whatever sound he wanted to pick up.

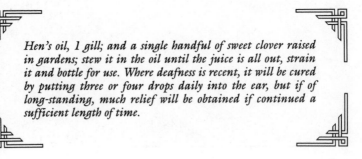

*Hen's oil, 1 gill; and a single handful of sweet clover raised in gardens; stew it in the oil until the juice is all out, strain it and bottle for use. Where deafness is recent, it will be cured by putting three or four drops daily into the ear, but if of long-standing, much relief will be obtained if continued a sufficient length of time.*

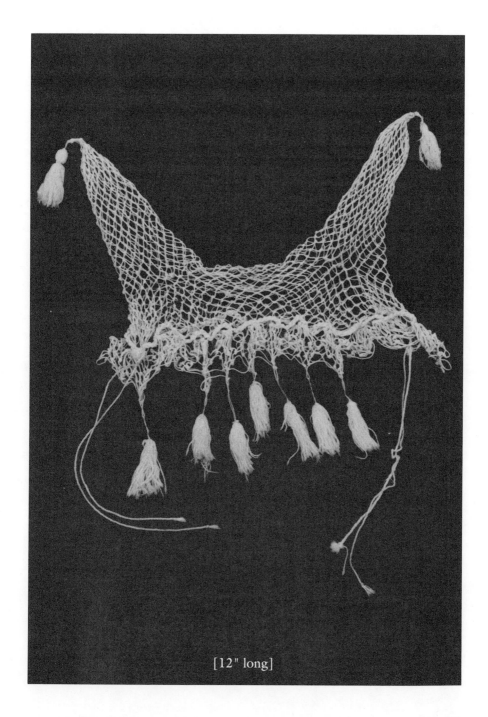

[12" long]

## HORSE EAR PROTECTOR

Before the advent of the automobile, the horse and buggy was the way to get from here to there on a daily basis. City-dwellers might have a fancy carriage for transportation on both business and pleasure trips. Country folk, if fortunate, had a no-nonsense work wagon as well as a buggy for Sunday best and social occasions. Whatever the rig, common sense demanded that it be treated properly—the wheels kept in good repair, harnesses well-oiled, tended, and mended, and any fancy work polished until it gleamed.

But the most important part of the rig was the horse, whether a high-stepping thoroughbred or a solid worker who was asked to go to both market on Saturday and church on Sunday. The animal deserved protection from the elements, good nourishment, and decent treatment in return for a lifetime of service.

This ear protector from the 1840s was undoubtedly designed with affection and concern. Made out of netting, knotted and with decorative tassels, it kept flies from pestering the horse's ears.

*To keep flies off horses, take two or three small handfuls of walnut leaves, on which pour two or three quarts of cold water; let it infuse one night, and pour the whole, the next morning, into a kettle and boil for one hour. When cold, it is fit for use. Moisten a sponge with it, and before the horse goes out of the stable, let those parts which are most irritable be smeared over with the liquor. Flies will not alight a moment on the spot to which this has been applied.*

[10"]

## ICE CREAM MAKER

This wondrous tin utensil was patented March 23, 1920, by the Liberty Can & Sign Company of Lancaster, Pennsylvania. There were smiles all around when it was taken out and put to use, and the event was apt to be a family affair. Mother began by mixing and cooking the ingredients—milk and cream, sugar, and flavorings. The results were poured into a metal canister, with a paddle or dasher inside which could be rotated by a crank on the outside. The canister was set into a larger bucket, and the space in between packed with chipped ice and rock salt. Then the real work began. Children took the first turns on the crank. When the mixture began to get mushy, fruit or nuts could be added. And when turning the crank required more muscle, father took over—until turning became impossible and everybody fell to eating.

*To make peach ice cream, 1 quart cream, ¼ peck peaches, 1½ cupfuls sugar and 1 lemon. Scald 1 quart thin cream and 1 cupful sugar; when cold, freeze to a mush, then add 1½ cupfuls peach pulp, mixed with ½ cupful sugar, and the juice of ½ lemon; finish freezing.*

[7½"]

## ICE CREAM SCOOP

This turn-of-the-century kitchen utensil could be used only upon completion of what was usually a family project—the making of ice cream. Mother would put together the ingredients—milk and cream, sugar, and flavorings like peaches, or strawberries, or chocolate. Father would pack the freezer all around with ice and rock salt, and the children would line up to help turn the crank, which turned the dasher, which turned Mother's concoction into everyone's dessert.

This ice cream disher, pictured in the 1902 Sears, Roebuck and Co. catalogue, was a companion to Shepard's Lightning Ice Cream Freezer. The scoop has two revolving knives which cut the cream loose. "By one-half turn of the button, the cream slips out a smooth and perfect cone." These sold for 10 to 14 cents, depending on the size.

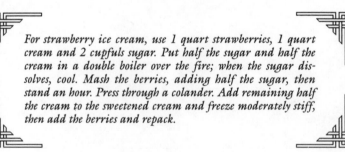

*For strawberry ice cream, use 1 quart strawberries, 1 quart cream and 2 cupfuls sugar. Put half the sugar and half the cream in a double boiler over the fire; when the sugar dissolves, cool. Mash the berries, adding half the sugar, then stand an hour. Press through a colander. Add remaining half the cream to the sweetened cream and freeze moderately stiff, then add the berries and repack.*

[4"]

## INK BLOTTER, PEN, QUILL SHARPENER

The simple task of writing a letter, keeping records, or making a list used to be a messy business indeed, because the hollow barrel end of a feather was split (to hold more ink), sharpened (with a "penknife"), and the point dipped in ink every few words. It couldn't be overfilled, or a drop would fall onto the paper. Writing neatly with such a pen required a light touch, so the point wouldn't catch on the paper and leave splotches of ink. The runny ink took a long time to dry, so the writing had to be blotted. One way to do that was to sprinkle sand on, allow it to soak up the moisture, then remove the sand by either blowing it away or picking up the paper and sliding it off. This writing set from the 1780s contains a pen, quill sharpener, and wooden bowl to hold blotting sand.

*To make good black ink, use ground logwood, one ounce; nutgall, three ounces; gum arabic, two ounces; sulphate of iron, one ounce; rain water, two quarts. Boil the water and wood together until the liquid is reduced one-half; then add the nutgalls coarsely bruised, and when nearly cold, the sulphate of iron and gum; stir it frequently for a few days, then let it settle; then pour it off and cork it up close in a glass bottle.*

[15"]

## INSECT SPRAYER

This object from the 1790s looks for all the world like a bellows for fanning the fire, and it operates on the same principle. You squeeze the handles, and by contraction and expansion, the bellows takes air in through one opening and expels it through another. In fact, the same kind of pumping device, only larger, forces air into the pipes of an organ. In this case the bellows, however, took in and gave off poison from an attached can. This early bugsprayer, although it could be used only on a small scale, helped rid plants of pests. It's 15 inches long.

*Experiments seem to indicate that all soft-bodied sucking insects are destroyed by contact with kerosene. Pure kerosene may be applied to the hardier trees in winter when they are not growing. For application to growing trees and foliage in summer, a mixture called kerosene emulsion is recommended, as follows: hard, soft or whale-oil soap, ½ pound; boiling soft water, 1 gallon; kerosene, 2 gallons. The soap is first dissolved in the water, then the kerosene is added and churned from five to ten minutes. Before using, this must be reduced with water from one-fourth to one-tenth its strength.*

[22"]

## IRON LAMP

Hand-forged of iron in the 1750s, rugged in the extreme, with a shallow, bowl-like, covered container, this lamp has a half-ball curving up over the top attached to a long rod with a hook so it could be hung from a nail or peg in the home, barn, privy, or ship. Its only decorative touch is a small bird, which serves as a handle to remove the lid.

This is a version of the very early "Betty" lamp, a wick-lighting device first brought to America on the *Mayflower.* Although the origin of the name is not certain, it may have been from the German word *besser,* meaning "better," or the Old English *bete,* meaning "to kindle a fire."

Filled with bad-smelling fish oil, it gave off not much more than a speck of light. There is no glass chimney to reflect the flame, just a tiny wick soaked in oil, animal fat, or candle ends. Attached is a half-ball of iron to extinguish the light. This was, most definitely, not a lamp to read by. It just made the night a bit less dark.

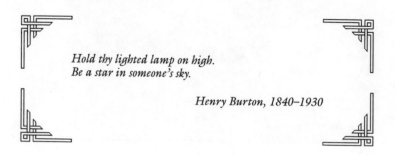

*Hold thy lighted lamp on high.*
*Be a star in someone's sky.*

*Henry Burton, 1840–1930*

74

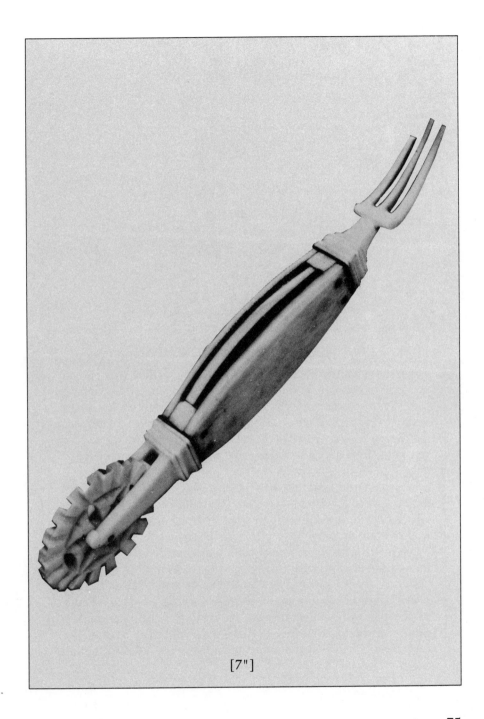

[7"]

75

## JAGGING WHEEL/PIE CRIMPER

When the whaling industry was at its height, sailors who signed on for a voyage were at the mercy of their captain and the fishing fates. Even the shortest trip lasted for several months. But if the hunting was poor, if whales were few and far between, if the barrels weren't filled with sperm oil, a ship would stay at sea for years at a time, its crew having not a glimpse of home or loved ones.

To pass the time, sailors took to carving whalebone and doing scrimshaw (etching with a jackknife or sail needle on a whale's tooth or bone). When they returned home, their work was lovingly presented as a gift from the sea.

While a whalebone corset busk was considered appropriate for a sweetheart or wife, a jagging wheel was a more restrained and very useful present for a mother or sister. Also called a wheel jigger or pie crimper, this 1820s jagging wheel is a hand-carved piece of whalebone, and its turning wheel was used to flute or crimp together the edge of a pie shell before baking. The cook used the fork at the opposite end to poke holes in the top crust.

[2½" h x 7" w]

## LADY'S SPITTOON

This item probably was not found in every home at the turn of the century, and it most assuredly was not an object for public display or use. Thought to be part of a washstand set, this example from the 1860s might have been used by a woman who had a cough or who had—perish the thought—been smoking, chewing tobacco, or taking snuff.

*Spitting is a filthy habit and annoys one in almost every quarter, indoors and out. No refined person will spit where ladies are present or in any public promenade; the habit is disgusting in the extreme, and one would almost wish that it could be checked in public by means of law.*

[18" l]

## LEATHER STRETCHER

Until fairly recently, a closetful of shoes of every sort was a luxury enjoyed by only the very wealthy. For the most part, people had two pairs—everyday and Sunday-best. Further, work shoes and boots, especially, were made not always for comfort but for duration, and rarely in a wide variety of sizes and widths. More often than not, the foot had to fit the shoe, not vice versa. A local repair shop was not available to make special adjustments for arthritic toes, bunions, or corns.

Both the iron tool manufactured in 1881, and the more modern wood-and-metal one from the 1920s provided comfort to a man or woman who owned a new pair of shoes or had a bunion. The knob end of the tool was placed inside the shoe at the appropriate place. The cobbler squeezed the grip and held it until the leather was stretched enough to accommodate the bunion. (It was not necessary to oil or wet the shoe before stretching it, but it never hurt to rub in a little neatsfoot oil to soften the leather.) The more up-to-date stretcher has a wing-nut mechanism on the handle, so it can be tightened to hold the tension.

*For corns, use a salve made of equal parts of roasted onions and soft soap; apply it hot. Or apply a sponge wet with a solution of pearlash. Wild turnip scraped and bound upon the corn, after the corn has been cut and made tender, will cure it in a short time.*

[9½" l x 2¾" w]

## LID HANDLE, ETC.

This small cast-iron item was well designed to serve three purposes in the busy kitchen of the 19th century, where the focus of much activity was the wood-burning cook stove. First, it is a lid handle. The small projection at one end fit into the recessed notch in the lid of each stove burner, which had to be lifted in order to feed the fire. Those lids were made up of concentric rings that nested into each other to make a flat surface. One ring at a time could be removed to expose a pot to more and more direct heat from the fire below.

Second, the object served as a trivet to hold hot pans just off the stove. Turn it over, and the jagged grid on the bottom reveals its third use—as a meat tenderizer.

*The cook stove may be kept in good order by a daily brushing or rubbing and by a thorough blacking and polishing once a week. To keep it clean, sprinkle a little salt over anything that is burning on the stove to remove the dirt by rubbing. Have at hand small sheets of sandpaper to remove whatever adheres.*

[15"]

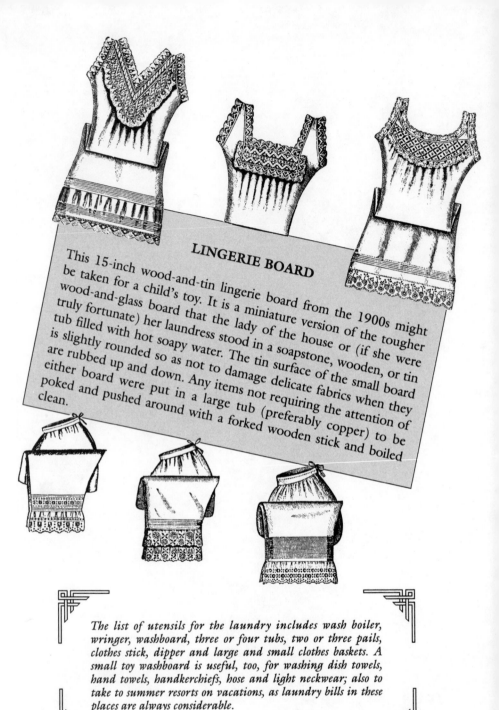

## LINGERIE BOARD

This 15-inch wood-and-tin lingerie board from the 1900s might be taken for a child's toy. It is a miniature version of the tougher wood-and-glass board that the lady of the house or (if she were truly fortunate) her laundress stood in a soapstone, wooden, or tin tub filled with hot soapy water. The tin surface of the small board is slightly rounded so as not to damage delicate fabrics when they are rubbed up and down. Any items not requiring the attention of either board were put in a large tub (preferably copper) to be poked and pushed around with a forked wooden stick and boiled clean.

*The list of utensils for the laundry includes wash boiler, wringer, washboard, three or four tubs, two or three pails, clothes stick, dipper and large and small clothes baskets. A small toy washboard is useful, too, for washing dish towels, hand towels, handkerchiefs, hose and light neckwear; also to take to summer resorts on vacations, as laundry bills in these places are always considerable.*

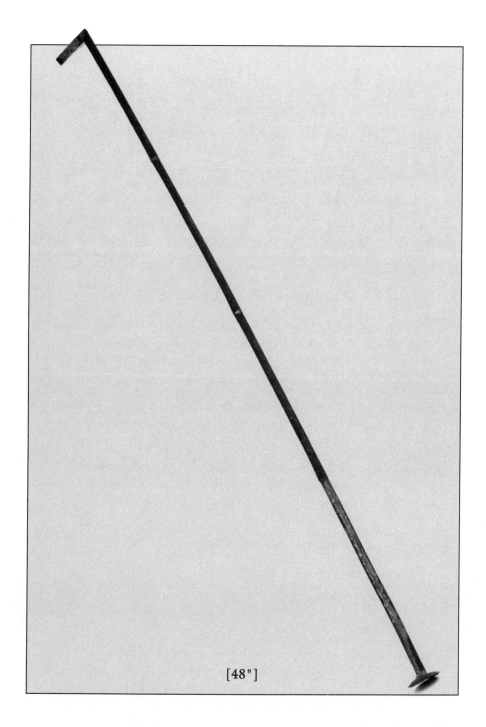

[48"]

85

## LONG CANDLE EXTINGUISHER

This tool from the 1770s has no moving parts or complicated purpose. It was designed with simplicity and plain common sense to
do one job that had to be done, and it likely was found in a church
or town meeting hall.

In public gathering places with high ceilings, candles, the sole
source of light, often were placed above the reach of people. This
four-foot-long tin candle extinguisher was used by placing the
small end near the flame and blowing through the other. Of
course, how those high-up candles were lighted in the first place is
another matter to ponder.

[4"]

87

## MATCH HOLDER

There was a time when every last source of heat and illumination in a house was activated not by flipping a switch or pushing a button, but by striking a match. Stick matches were kept all around, some in a tin holder on the kitchen wall next to the stove, and others in more decorative—more imaginative—containers throughout the house. This is a cast-iron, log-shaped holder from the 1880s that keeps the stick matches safely tucked away and provides access by way of a sharp-beaked bird on a spring. When the woodpecker's tail end is lifted, his extremely sharp beak dips into the holder, spears the match stick, and lifts it out.

[5"]

## MECHANICAL PENCIL

Although most folks are used to having a ballpoint or felt-tipped pen in every pocket or purse, being able to carry ink around is a fairly recent luxury. In the days when using a pen required a nearby inkwell, the only writing instrument that could be carried outside the home was a pencil. But that proved both messy and impractical, because when the tip didn't break—which happened more often than not—the soft lead left marks on clothing. Thus the 1870s gave birth to the first of the mechanical pencils. The fine point of lead was twisted out as needed and retracted when not in use, and there were spare lengths of lead inside the case.

[6" dia.]

## MORTAR AND PESTLE

Before medicines and toiletries were available commercially, people used all sorts of recipes to make everything from pills and elixers to perfumed waters and rouge. Utensils required for compounding home remedies were found in most households, including druggist's scales, a measuring glass, a small spatula, glass rods for stirring, and a double boiler for melting and mixing liquids. And because so many home cures called for infusions made from steeping powdered ingredients, the raw materials first had to be pulverized and thoroughly mixed, making a mortar and pestle absolutely vital for the home apothecary. This wooden mortar with its pestle is from the 1780s. The mortar is hewn from a single piece of wood and has a concave bottom inside like a bowl. The ingredients are placed in it and crushed into powder with the rounded end of the pestle.

*To make Composition Powder—Bayberry bark 2 lbs; hemlock bark 1 lb; ginger root 1 lb; cayenne pepper 2 oz; cloves 2 oz; all finely pulverized in a mortar and well mixed. Dose—one-half of a teaspoon of it and a spoon of sugar; put them into a teacup and pour it half full of boiling water; let it stand a few minutes and fill the cup with milk. Drink freely. This is a valuable medicine, and may be safely employed in all cases. It is good for pain in the stomach and bowels and to remove all obstructions caused by cold. A few doses, the patient being in bed with steaming stone at the feet, will cure a bad cold and often throw off disease in its first stages.*

[5"]

93

## NUTMEG GRATER

Cooking with spices nowadays requires nothing more than shaking them out of a little jar or tin. But in the days before supermarkets, spices were available only in their natural form. If you wanted to cook with a certain spice, or use it to make perfumed water, you first had to grind it, break it, or pulverize it.

The nutmeg tree, an evergreen, produces two spices. The aromatic nutmeg is from the inner seed, and mace comes from the covering that separates the seed from its outer husk. This tiny tin grater has a receptacle on top and a handle that turns a perforated cylinder.

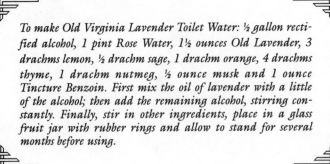

*To make Old Virginia Lavender Toilet Water: ½ gallon rectified alcohol, 1 pint Rose Water, 1½ ounces Old Lavender, 3 drachms lemon, ½ drachm sage, 1 drachm orange, 4 drachms thyme, 1 drachm nutmeg, ½ ounce musk and 1 ounce Tincture Benzoin. First mix the oil of lavender with a little of the alcohol; then add the remaining alcohol, stirring constantly. Finally, stir in other ingredients, place in a glass fruit jar with rubber rings and allow to stand for several months before using.*

[small, 6½" d; medium, 10½" d; large, 13½" d]

Table of Liquid Measure Equivalents

| | |
|---|---|
| 4 gills | 1 pint |
| 2 pints | 1 quart |
| 4 quarts | 1 gallon |
| 31½ gallons | 1 barrel |
| 2 barrels | 1 hogshead |
| 2 hogsheads | 1 pipe or butt |
| 2 pipes | 1 tun |
| 282 cu. inches | a beer gallon |
| 36 beer gallons | 1 barrel |

[ANSWER ON PAGE 98]

[4" h x 3" w]

## PANTRY BOXES AND MEASURES

Buying food today is dramatically different from what it was less than a hundred years ago. Shoppers go into the market empty-handed and come out carrying bundles of bundles—paper and plastic bags full of other paper and plastic bags. Almost all food-stuffs are prepackaged, preboxed, prebottled, and preweighed, and containers are seldom, if ever, used a second time.

It wasn't always so. In the first place, people didn't go market-ing on a whim. They shopped once a week—ordinarily on payday. Staples like flour and sugar were bought in quantity enough to last a month—or the winter, if weather was severe. Moreover, most goods were not packaged. Stores were lined with bins and jars, tins and barrels. Cornmeal, peas, beans, crackers, oats, wheat, spices—all were bought by the measure, and many were carried away in a container that came from home in the first place. You took a tin milk jug to the store and had it filled. You put a loaf of bread into your own net or cloth bag. You might, of course, buy a barrel of apples to put down for the winter. Goods were measured out with a pint tin, a pound scale or round wooden boxes of vary-ing sizes, like these. Made of bentwood, the boxes were often imprinted with a stamp to verify their size and make sure the cus-tomer didn't get short-changed on a purchase. These are grain measures from a store, each made to a standard capacity.

[6"]

## PAP BOAT/INVALID FEEDER

Until well into the 20th century, most homes contained three generations of a family. Children seldom left their parents' house until they married, and grandparents were part of the household all their lives. Elderly relatives were not sent to nursing homes, because there were no such places. In the country, if a house got too small an ell was added—and another and another. In the city, where that option wasn't available, a little girl might very well have shared a room with her grandmother.

A small porcelain object like this one from the 1880s was found in many homes, because it was used at both ends of life and, occasionally, in the middle. Depending on the need at a given moment, it was an infant feeder or invalid feeder. Those of a slightly elongated shape were called "pap boats." They permitted someone lying down to take food or medicine without spilling it.

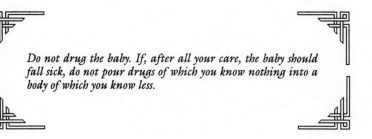

*Do not drug the baby. If, after all your care, the baby should fall sick, do not pour drugs of which you know nothing into a body of which you know less.*

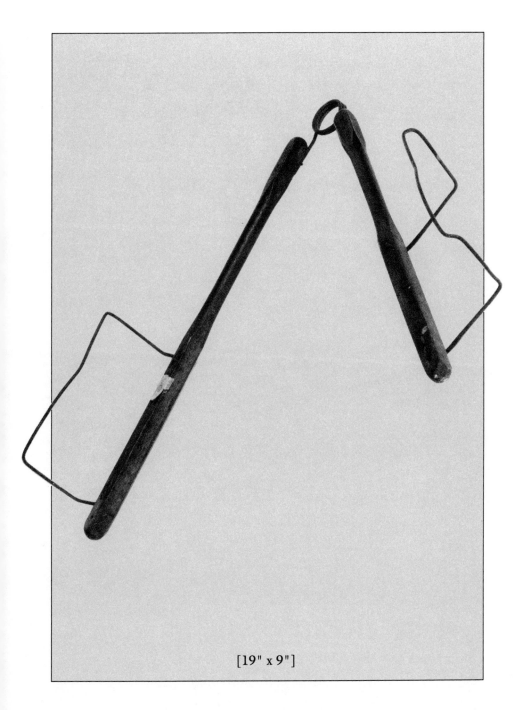

[19" x 9"]

HOME MADE
PIES
A SPECIALTY

## PIE LIFTER

Although this simple item from the 1870s could easily be replaced by squares of padded cloth or folded dishtowels, it did save the cook from having to reach all the way into a hot oven. The two wooden arms are joined at one end by a coil of metal that acts as a spring. At the opposite end are two thin pieces of metal that serve to keep the hot pan from sliding off the holder as the cook lifts the pie.

[12" l x 19" d]

## "PIG" HOT WATER BOTTLE

Because of its shape, this original hot water bottle was affectionately called a "pig" by the British. It could be found in most rooms of the house in the days before central heating. Made of salt-glazed country pottery, this one has a hole in the top with a cork stopper. It was filled with boiling water, and the thick ceramic held the heat for a good amount of time. Place it under the covers at the foot of a bed for a while. Then pull it out by grasping the snout end. Or set it on the floor in front of a chair, where it will serve to warm someone's feet.

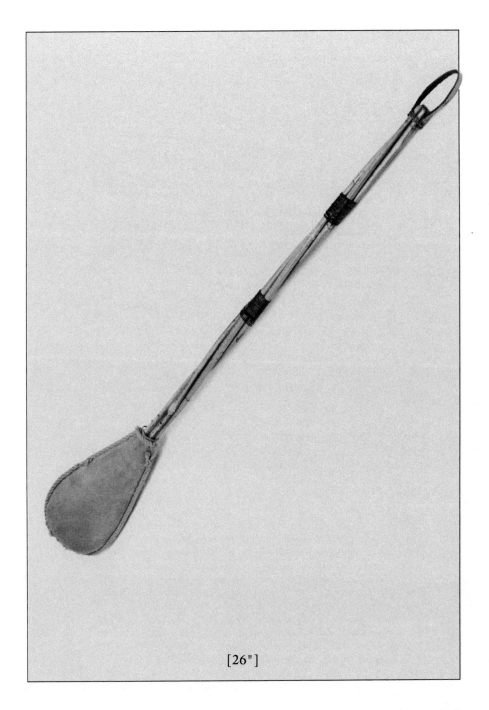

[26"]

## PILLOW FLUFFER

The use of natural materials to equip a bedroom included mattresses made of hair, moss, dried corn husks, or—in the more prosperous homes—goose or duck feathers. The finest of these—the down—was used to stuff comforters and pillows. When feathers were put loose into a stitched-up piece of ticking, they shifted and settled to the side or end after a certain amount of use. While it was common to clean feather beds and pillows every two or three years by taking them apart, scrubbing the ticking and purifying the feathers, freshening and fluffing were done far more frequently. Two people picked up the mattress by one end, shook all the feathers to the other, did the other way around, then repeated that action from side to side, loosening and distributing the feathers evenly throughout the ticking. The pillow got the same treatment, and if possible, both were taken outdoors for a good airing. This feather-pillow fluffer was used to put a little extra muscle into the cleaning job. Hold the pillow in one hand or hang it from a line, and beat it, removing dust and stirring up the feathers.

*Feather pillows may be washed without removing the feathers by boiling them in borax water to which a small quantity of ammonia has been added. Use half a teacupful of borax to a boilerful of water and add a tablespoon of ammonia. Boil 15 or 20 minutes. After removing, scrub the tick, if badly stained, by laying it on a washboard and applying suds with a stiff brush. Rinse in two or three waters and hang on a line in a shady place to dry. Shake the pillow and change ends two or three times a day. It takes a long time to dry. This process makes the feathers light, flaky and sweet-smelling.*

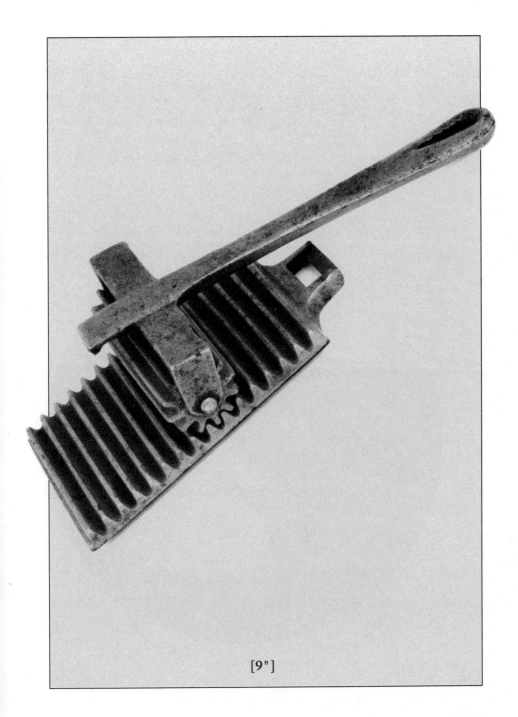

[9"]

107

## PLEATING IRON

Ironing must have been an interminable chore—perhaps second only to doing the wash. Women's clothes contained yards of material. There were overskirts and underskirts, bodices with fancy work—laces and ribbons and pleats every which way. Men's clothing required laborious ironing, too. There was no such thing as "permanent press." The irons were no great shakes, either, all of them heavy cast-iron things that had to be set on the stove until hot. But this nine-inch pleating iron from the 1860s may have been one small blessing to the laundress faced with mountains of intricate dresses. It, too, was heated on the stove. Then it was placed just so in a garment and the top lowered to form many pleats at once.

[8" x 9"]

## ROOM HEATER

Before the days of central heating, homeowners relied on fireplaces and wood stoves to keep the chill away, and this object shows how they coaxed a little extra warmth out of a simple kerosene lamp.

The Falls Heater, manufactured in Boston in 1892, was advertised as "the simplest, cheapest, and cleanest method of heating rooms. By application of a scientific principle, the heat of the lamp is increased and distributed evenly throughout the room."

The tin heater, like a small stovepipe in appearance, was suspended over the lamp, where it caught the warm air rising and diverted it in opposite directions by way of two small vents. In addition to providing some heat in the room, it could be used as a hand warmer.

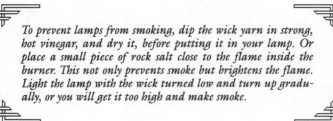

*To prevent lamps from smoking, dip the wick yarn in strong, hot vinegar, and dry it, before putting it in your lamp. Or place a small piece of rock salt close to the flame inside the burner. This not only prevents smoke but brightens the flame. Light the lamp with the wick turned low and turn up gradually, or you will get it too high and make smoke.*

[48"]

## RUG BEATER

This fancily woven tool, circa 1910, was an oft-used, entirely necessary piece of household dry-cleaning equipment. The process required no chemicals or detergents but provided, instead, a great deal of exercise and a way to let off steam. Perhaps for those reasons alone, it was common to assign rug beating to a child.

All the floor coverings in the home were given a good going-over during spring cleaning, but the smaller ones were taken up more often. Rag rugs, braided and hooked ones alike, were draped over the clothesline to take their beating. Larger rugs were put on the ground, beaten until the dust rose to the surface, and then swept. Whenever possible, this task was done on a day filled with sunshine and a good breeze to freshen the rugs.

*The most efficient agent to sterilize dust, by killing germs that it contains, is direct sunlight. Like many other things that are plentiful and free, sunlight is not appreciated at its true value. In the presence of sunshine, dust is rendered harmless. Hence choose furnishings that sunlight will not harm and admit it freely to all parts of the house.*

*Rugs should be removed from the room and thoroughly cleaned before they are returned to the floor. Hang them on a line and beat them with a carpet beater. If rugs must be beaten indoors, lay a damp cloth over them. Or lay a rug on a clean floor and sprinkle table salt over it. Sweep it hard with a broom until it is clean.*

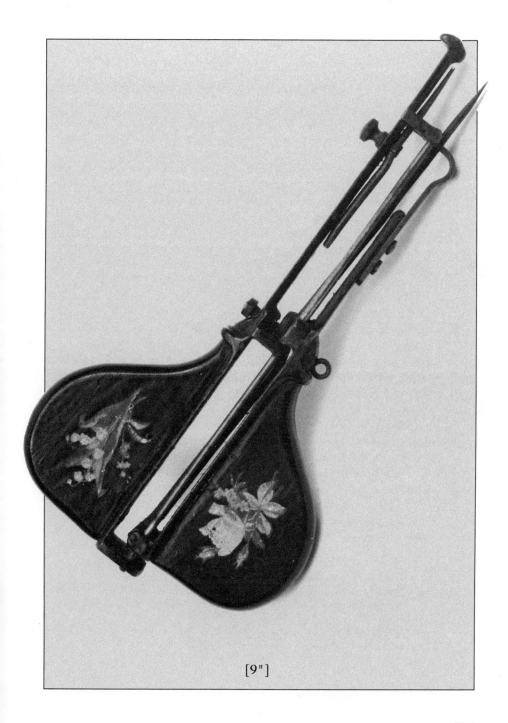

[9"]

113

## RUG-HOOKING TOOL

The art of rug hooking, which dates back to ancient Egypt, was tremendously popular in 18th- and 19th-century America, primarily because it permitted people to take otherwise useless bits of material and turn them into something both attractive and useful. Most of our antique hooked rugs come from that period, and although materials and designs have since become somewhat more sophisticated, the basic art remains the same. It's a craft that is time-consuming and requires patience.

Simply put, strips of wool or yarn (even rope or string in the old days) are pushed through a hole in a piece of sturdy backing, then brought up through another hole, forming a "U" shape of the filler material. Today that backing is canvas or burlap, often with a design already stamped on it. A hundred years ago designs were free-form, and the backing might have been a potato sack. Itinerant peddlers sold templates for a wide variety of rug designs. Traditionally, hooking involves both hands: one of them holding a hooking tool and forcing the material down through the hole; the other, underneath the backing, where it pushes the yarn or wool back up.

This is a semi-automated "Jewel" hooking tool, patented in January 1886. It works almost like a sewing machine. Wool yarn is fed through the large needle, and the tool is used almost like a punch—to jab the material into the backing (from the front), making loops on the top surface. It was a faster way to hook a rug but did not permit much in the way of curved lines, as in the popular floral designs of antique rugs.

*Never shake a hooked rug to remove soil. The wool and backing will weaken and tear where it is held by the hand. Lie the rug flat and wash with mild soap. Or allow to be snowed on, which will purify the rug and brighten the colors. Dry thoroughly, but not in hot sun. The burlap is the most easily damaged part of the rug and may rot due to dampness. Store rugs flat. If they must be rolled, make sure the design is on the outside of the roll.*

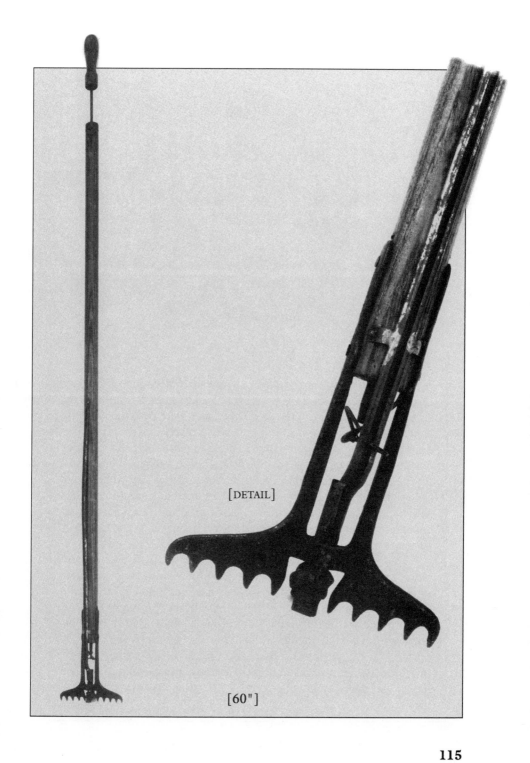

[DETAIL]

[60"]

## RUG TACKER

In 19th-century homes—before anyone dreamed of wall-to-wall carpeting—wood floors were covered only here and there, by either woven carpets or rag rugs. Large rugs were used primarily in parlors and dining rooms, to cover worn places and knots in the floor and prevent a draft from coming up between the boards. Small rugs were placed next to beds and in the kitchen. These scatter rugs weren't attached to the floor, because they had to be taken up, cleaned, and beaten frequently. But something had to be done about the larger, room-sized rugs. They were apt to wrinkle, move about when furniture was shifted, and curl up at the edges and corners.

This tool, made in 1901, has a long, narrow shaft on the handle, a forklike piece of metal at the bottom, and a simple geared turning mechanism near the base. It's five feet long. Fill the shaft with tacks, their heads against the handle. Place the bottom of the tool at the edge of the carpet, holding it in place. Pull the handle up, then push it down with force. A tack goes down the shaft, is turned point-down by the gear, and driven through the rug into the floor. By simply walking along, pulling and pushing, you can tack down a carpet in short order—no muss, no fuss, no aching back.

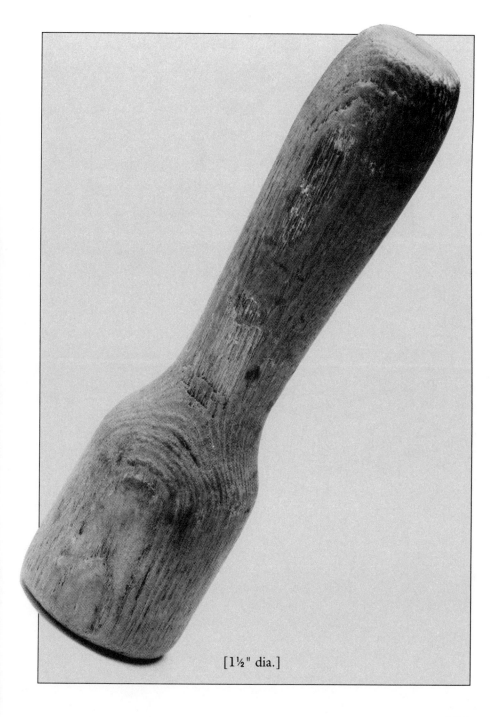

[1½" dia.]

## SAUSAGE PLUNGER

This piece of wood that traveled from the lake to the kitchen is another example of what it meant to "make do." Before sausage casings were stuffed automatically, the sausage maker used an implement created by simply cutting the handle off a broken rowing oar. It could also be used, one supposes, as a pestle for grinding and mixing medicinal concoctions or toilet preparations—including powders, essences, and rouges.

[7"]

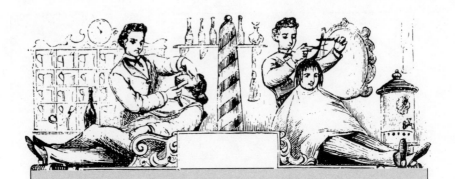

## SCUTTLE SHAVING MUG

In the days before hot running water, electricity, and aerosol cans, some version of this mug (the one here was made sometime between 1890 and 1910) was part of most every gentleman's toilette. There were round ones and oblong, some with floral decoration, some with hunt scenes, and others with advertising slogans. But the basic design was the same, as was the function.

The deep part was filled with hot water from the kettle. A piece of soap was placed in the dishlike space at the top. In addition, the whole process required a soft-bristled brush and straight-edged razor.

*Those who shave should be careful to do so every morning. Nothing looks worse than a stubbly beard. The style of the hair on the face should be governed by the character of the face. Whatever the style be, the great point is to keep it well brushed and trimmed and to avoid any appearance of wildness or inattention. As dust is sure to accumulate in a full beard, it is very easy to suffer it to become objectionable.*

*However, there are many who believe that the person who invented razors libeled nature and added fresh misery to the days of man. "Ah," said Diogenes, who would never consent to be shaved, "would you insinuate that Nature had done better to make you a woman than a man?"*

[12"]

The great enemy of leather—especially patent leather—is heat. Extreme heat tends to rob the leather of its vitality and causes it to break and crack. Damp shoes should never be placed near a stove to dry, since if heated enough to give off the characteristic odor of leather, they may be singed and ruined. To dry shoes, place them on their sides in a warm room, in a draught of dry air if possible, but not near a fire. Or heat bran or sand and fill two stockings, tying the tops tightly. Put the shoes on these as on shoe trees.

[ANSWER ON PAGE 124]

122

[12½" l]

## SHOE ANVIL

This is a portable iron bench-top anvil from the early 1900s, used in the repair and resoling of shoes and boots. Economy of design and practicality are its hallmarks. If a cobbler had to work on both women's and men's footwear, why have two separate forms, each one requiring its own sturdy base? Better to create one solid piece of equipment to meet both needs.

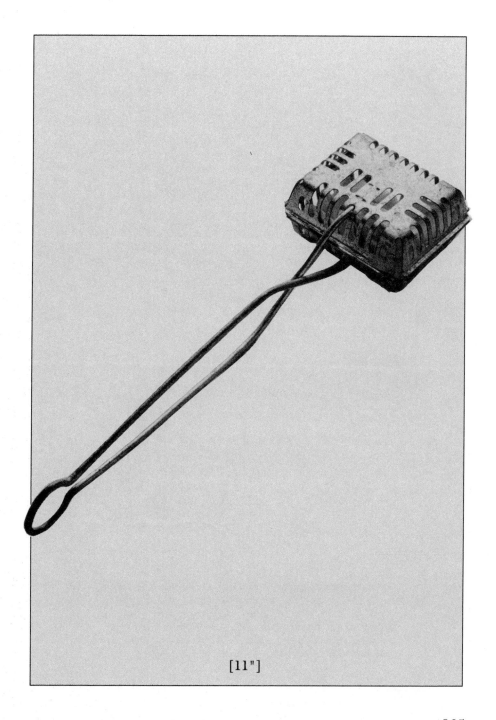

[11"]

## SOAP SAVER

This simple, metal, basketlike device is further evidence that to waste anything was unforgivable in a day when home-cleaning and personal-care items were hard to come by and soap was handmade. When a cake of soap got too small to grasp, it was put into a holder like this example from the 1940s, along with other leftover bits of soap. The soap saver—usually hung by the kitchen sink or washtub—was then swished through the water to make suds. Before the development of detergents, the dishpan of soapy water could be thrown out the back door onto the garden. Plants loved the fat in the soap and bits of food in the water.

*To make soft soap, mix in a kettle or wash boiler 8 pounds of melted grease with 1½ pailfuls of strong lye that will float a fresh egg. Bring to a boil, pour into the soap barrel and thin with weak lye obtained by leaching wood ashes. Place the barrel out of doors in warm water or in a warm place. The soap should be ready for use in a few days.*

126

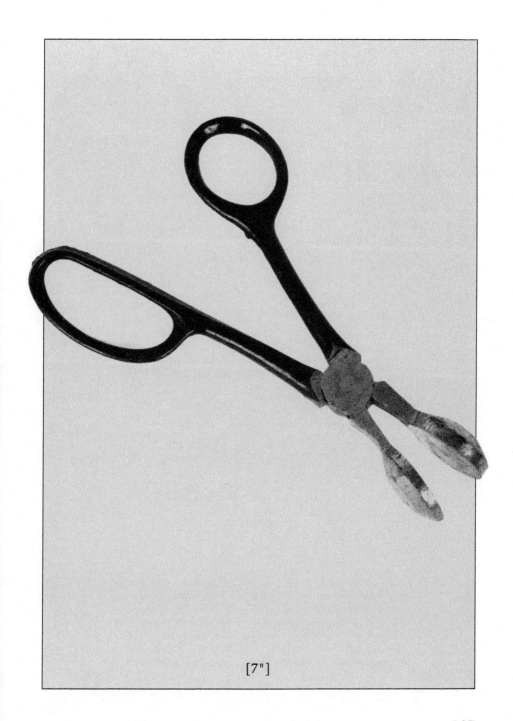

[7"]

## SPIT CURLER

At first glance this looks like a mold used to make shot for guns. But the insides of the two metal parts that squeeze together are flat, so there's not room for much between them. Rather than being useful in the barn or workshop, this implement was part of a lady's toilette.

First, it had to be heated over a stove, alcohol burner, or gas jet heater. While the metal was getting hot, the woman twisted some locks of hair—on her forehead, perhaps—into a tight curl. Then she picked up this spit curler, put the round ends on either side of that curl and squeezed the handle. Made in 1890, it's a handy seven inches long.

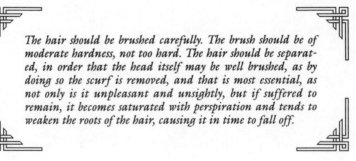

*The hair should be brushed carefully. The brush should be of moderate hardness, not too hard. The hair should be separated, in order that the head itself may be well brushed, as by doing so the scurf is removed, and that is most essential, as not only is it unpleasant and unsightly, but if suffered to remain, it becomes saturated with perspiration and tends to weaken the roots of the hair, causing it in time to fall off.*

[1"]

## STANHOPES

It's hard to imagine a time when there wasn't a television in the house, a radio in the car, and a movie theater down the road. Entertainment is so close at hand and automatic that we're at a loss for things to do when the power goes off. But before that "power" was discovered, there were diversions of another sort. People read aloud, played games, put on dramas and tableaux, made music, danced to it, and turned common work projects into social gatherings—like quilting bees and corn-husking bees, barn-raisings and sugaring-off parties. And every now and then, someone came up with an invention that was truly astonishing.

Charles Stanhope was a British politician and inventor who, among other things, opposed England's war with the American colonies. He was born in 1753 and died in 1816. His scientific experiments produced fireproof stucco, the first calculating machine, and machines for printing and stereotyping. The Stanhope press and Stanhope lens are named after him. So are these inch-long objects—small bits of entertainment made in 1896 and called, simply, Stanhopes. Carved of ivory, they contain a tiny lens and either pictures or words. Hold them up to the light. Look in. The "binoculars" contain "Scenes of Monte Carlo." In the single Stanhope you can read The Lord's Prayer in its entirety.

*Three things are to be borne in mind while getting up amusements for a party. First, to get up an entertainment that as many as possible can partake in, for participation is part of enjoyment. Second, that in the entertainment there shall be nothing to which there can be any objection, or which shall leave unpleasant memories. Third, that the real object of the amusement be gained, namely, that all shall be amused.*

[9"]

## SUGAR NIPPERS

Before trade with the West Indies became well established, sugar was a precious item enjoyed in quantity by only the very well-to-do. It was not sold loose, in bags, but in cone-shaped loaves, usually about three feet high and weighing 15 pounds. The wealthiest shoppers might buy an entire loaf. Others bought sugar by the pound or ounce, and the grocer snipped or nipped the requested amount off the loaf using sugar nippers that worked like pliers but had sharp blades at the end of rounded arms.

[19" h x 8" w]

## SWIFT

Winding homespun wool yarn was, for the most part, a two-person job, involving a mother and daughter, husband and wife, or perhaps a young woman and her betrothed. The team sat close together, one of them with hands outstretched holding a hank of yarn. The partner took one end of that yarn and, unlooping it from the other's hands, began winding it into a ball.

In the 18th and 19th centuries, the husband or sweetheart who provided that second pair of hands might very well have been off on a whaling voyage for months or years at a time. Clever sailors devised a substitute for the women to use—a "swift," or wool winder, and they made it out of the only materials available to them—ivory, baleen, or whalebone. The swift was the most complicated piece of scrimshaw, usually containing more than 100 pieces fastened together with silver, brass, or copper rivets and even ribbon. The reel, which opened like a folding gate, revolved on a center spindle. A fastener slid up that spindle to hold the swift open—much like an umbrella. Sometimes an ambitious sailor mounted the swift on a box containing drawers for sewing or knitting supplies. Swifts were used from 1800 to 1850.

134

[3¾"]

## TATTING SHUTTLE

In the Victorian era, when almost every surface in the parlor, dining room, and bedroom was decorated with a doily or lacy tidy and when the backs and arms of chairs were protected with antimacassars (originally named this because they prevented soil marks from a then-popular hair oil imported from Macassar on the island of Celebes in Indonesia), women spent no small amount of time doing handwork. Embroidery was a popular form of needlework. However, the fanciest items were produced not with a needle but with this tiny object—and clever hands.

Tatting, an early relation of macramé, is knotted lace made from cotton or linen thread wound through a shuttle. The shuttle then is worked this way and that, over and under, to fashion the knots and pattern. The shuttle is held in one hand, and the piece takes shape on the other, the thread wound around the fingers.

Thought to have its beginnings in seamen's knots, tatting employs just one of these—the half hitch—worked in pairs. By using different threads and varying the spacing, tension, and direction of pairs of knots, the most delicate and intricate bits of lacework were created. This is a silver tatting shuttle, made in 1900.

[¾" dia.]

## THIMBLE HOLDER

When women spent a great deal of time both creating the family's wardrobe and keeping it in good repair, they couldn't be too far from their favorite thimble. This handsome silver case from the 1880s was worn around the neck, with the thimble inside ready for use whenever the need arose, which it did with astonishing regularity.

In the regular routine of weekly work, it is a good idea to set aside a day in which all garments may be mended, altered or made over. The day following ironing day is usually most convenient. The woman who has a room in her house which can be set apart solely as a sewing room knows not how to value her blessing.

138

[8" h x 12½" d]

139

## THREAD AND NEEDLE HOLDER

Just because something was quite practical in its purpose was no reason for it to be unpleasing to the eye. This late-19th-century brass sewing aid dealt with the most everyday of chores but in an altogether delightful manner. The whole thing turns on its base. There are several spindles, each of which may be removed by the twist of a screw at the bottom, and there is a rounded pad on the top. That pad is a pin cushion. Each spindle holds several spools of thread, and the seamstress of the house could easily find the appropriate color and snip off the proper length.

*When mending day comes around, the busy woman may find it hard to recall the things that need attention. Her memory may be assisted by recording in a small blank book the articles that have been out of service through rents or tears or lack of buttons and the like. On mending day the memorandum may be consulted.*

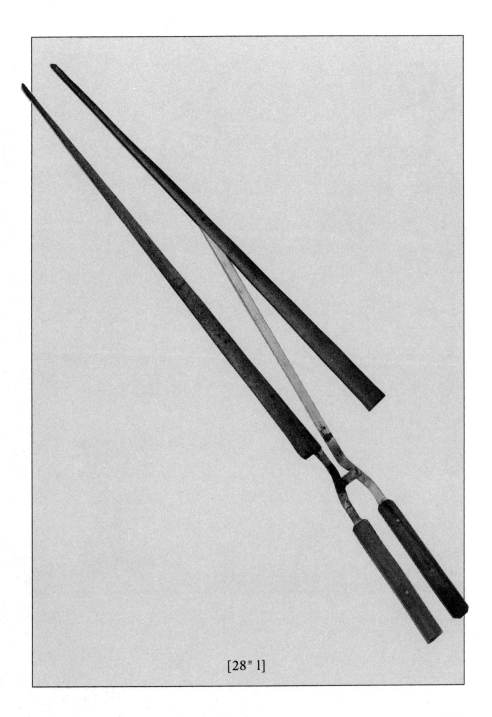

[28" l]

## TIE FORMER

Before the days of dry cleaning, of gentle cycles and permanent press, clothing of all description was boiled clean in a copper tub and run through a wringer. Although everything came out clean as a whistle, it was sometimes necessary to stretch and reshape a piece of clothing that had, literally, gone through the wringer. The wooden pieces of this 1930s tie former went into a man's tie—beginning at the wide, pointed end—and the handles were separated until the tie took on the desired shape. It then dried on the form, which could also be used to shape sleeves.

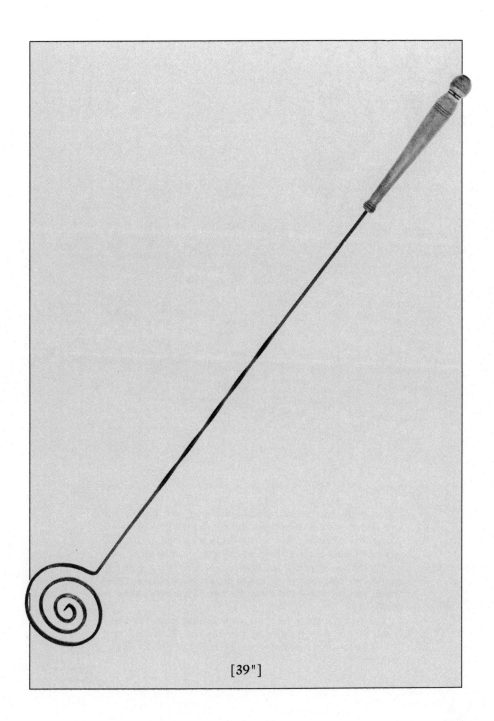

[39"]

## TOASTER

This object, with a wooden handle at one end and spiral of iron at the other, was put to use at either the fireplace or wood stove. The blacksmith twisted the long piece of iron in the making, giving it greater strength.

In the days before preservatives and good storage methods, bread lost its freshness quickly, and it was discovered that stale bread could be made quite tasty when toasted. In the 1790s, a family member would put a piece on the iron part of this wooden-handled implement and hold it over an open fire until brown, then turn the bread and toast the other side.

*Look carefully to the stale bread remains of each day. Keep a wire basket, set in a tin pan in the pantry, to receive all scraps left on plates. Never put them in a covered jar or pail; they will mold. Save all soft inside parts of a loaf to be used for croutons or croustades, slices or cubes for toast and soft scraps for meat and fish dressings, puddings, omelets, griddle-cakes and the numerous dishes for which stale bread may be utilized.*

*To make toast, trim the crust from stale slices and move them over a clear, red fire for two minutes. Then turn and let all the moisture be drawn out of the bread. Butter immediately.*

[8"]

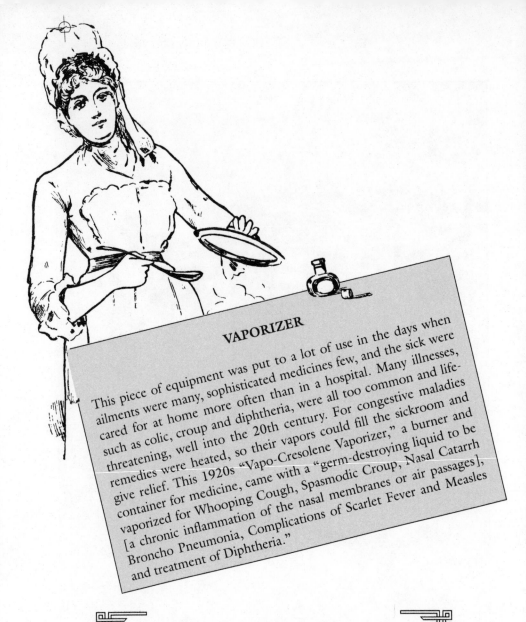

## VAPORIZER

This piece of equipment was put to a lot of use in the days when ailments were many, sophisticated medicines few, and the sick were cared for at home more often than in a hospital. Many illnesses, such as colic, croup and diphtheria, were all too common and life-threatening, well into the 20th century. For congestive maladies remedies were heated, so their vapors could fill the sickroom and give relief. This 1920s "Vapo-Cresolene Vaporizer," a burner and container for medicine, came with a "germ-destroying liquid to be vaporized for Whooping Cough, Spasmodic Croup, Nasal Catarrh [a chronic inflammation of the nasal membranes or air passages], Broncho Pneumonia, Complications of Scarlet Fever and Measles and treatment of Diphtheria."

*To make catarrh snuff, use Scotch snuff, 1 oz; chloride of lime, dried and pulverized, in rounding tea-spoon; mix, and bottle, corking tightly. The snuff has a tendency to aid the secretion from the parts; and the chloride corrects unpleasant fetor.*

146

[30" sq]

## VEGETABLE CHOPPER

If the discovery of electricity changed the face of the earth, it also altered the look of every tool, appliance, or piece of machinery that it touched. You cannot see the workings of any electrical device. All of that mysterious stuff is placed within a housing of some sort—out of reach, out of sight, and beyond control. This 1905 chopper shows the way things used to be. The works, blades, gears are there for all the world to behold. People had to know how things worked, so they could be fixed when they didn't. This "Mechanical Vegetable Chopper" was simple. It was durable. And it worked. Any sort of vegetable can be placed in the bin, which has a wooden floor. Turn the handle and the bin rotates. The blade is driven up and down by means of a simple ratchet mechanism. The faster the turn, the quicker the action. This was thought to be especially effective when chopping ingredients for piccalilli.

[8" x 9½"]

## WALL-LAMP BRACKET

This handsome metal bracket, made in 1878, is one more item made obsolete by the introduction of electricity to the home. In many a well-appointed Victorian residence, however, it could be found behind a chair, near a bed, or beside a doorway leading from one room to another. When kerosene lamps provided indoor lighting, they were often carried from place to place—up dark winding stairways, along narrow halls, into dark rooms or out to a cold backyard privy. Sooner or later, the lamp had to be set down somewhere. It was useful to have a fold-down wall bracket like this one handy, particularly if there was no room or other need for a table. When the person—and the lamp—moved on, the bracket could be folded up out of the way.

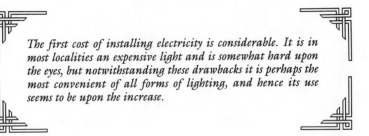

*The first cost of installing electricity is considerable. It is in most localities an expensive light and is somewhat hard upon the eyes, but notwithstanding these drawbacks it is perhaps the most convenient of all forms of lighting, and hence its use seems to be upon the increase.*

[24"]

*If it is possible to avoid it, never allow the woman who washes the clothes, and thus becomes warm and sweaty, to hang them out, and especially ought this to be regarded in the winter or windy weather. Many consumptions are brought on by these frequently repeated colds, in this way. It works upon the principle that two thin shoes make one cold, two colds an attack of bronchitis, two attacks of bronchitis one consumption—the end, a coffin.*

## WOODEN WRINGER

Doing the wash a day-long effort that required a sturdy back, strong arms, plenty of hot water, and a few basic pieces of equipment. Everything in the home—from dishtowels to bedsheets—had to be washed, wrung out, and hung on the line to dry. Until the invention of this wooden apparatus, the strenuous chore of wringing was done by hand. "Hall's Little Wringer," 1870, was one of the earliest devices for squeezing water out of clothes. Unlike later models, there are no metal parts, no removable, replaceable rollers, and no clamps by which the wringer could be fastened to a washtub. The laundress fed the wet wash between the two-foot wooden rollers, held the wringer in place with one hand, and turned the handle with the other. It wasn't easy—just easier.